Jacob Reinegg

F. A. Marschall von Bie

A GENERAL, HISTORICAL, AND TOPOGRAPHICAL

DESCRIPTION

of

Mount Caucasus

WITH A CATALOGUE OF PLANTS INDIGENOUS TO THE COUNTRY

In two volumes

Translated

by

CHARLES WILKINSON

Volume II

Elibron Classics
www.elibron.com

Elibron Classics series.

ISBN 1-4021-6253-7 (paperback)
ISBN 1-4021-2799-5 (hardcover)

This Elibron Classics Replica Edition is an unabridged facsimile of the edition published in 1807, London.

GENERAL, HISTORICAL, AND TOPOGRAPHICAL

DESCRIPTION

OF

Mount Caucasus.

WITH A CATALOGUE OF PLANTS INDIGENOUS TO THE COUNTRY.

IN TWO VOLUMES.

TRANSLATED FROM THE WORKS OF

DR. REINEGGS AND MARSHAL BIEBERSTEIN,

BY CHARLES WILKINSON.

William Couling, 1817.

VOL. II.

LONDON:

PRINTED FOR C. TAYLOR, HATTON-GARDEN; W. MILLER, ALBEMARLE STREET; A. COLLINS, 'CHANGE ALLEY; AND DARTON AND HARVEY, GRACECHURCH-STREET.

1807.

Printed by W. Darton, and J. & J. Harvey, Gracechurch-street.

A

GENERAL, HISTORICAL, AND TOPOGRAPHICAL

DESCRIPTION

OF

CAUCASUS.

GEORGIA.

The Provinces of Kaget and Kisik.

IF hiſtory be worthy of credit, the whole tract of country, ſtretching from the left bank of the Kurr to the Caſpian, was formerly ſtyled Albania. But the name would ſcarcely be remembered at preſent, if the Zaar Herakleus did not take it as one of his titles. I ſhall only mention thoſe provinces which are under his ſovereignty, and reckoned with Georgia.

At the conflux of the rivers Arakui and Kurr, where the ancient boundary between Albania and Armenia was fixed, ranges of mountains, of unequal elevation, are ſeen riſing towards the S. E. They are ſeparated by fertile valleys, and gradually ſink into an immeaſurable plain, where the Kurr winds its courſe in a ſouth-eaſtern direction.

Hallawar, a ſmall city, and the fortreſs of Medegh, ſtand in this plain, on the left bank of the Kurr; and are connected with Teflis by means of a bridge over the river. Hence they are conſidered as its suburbs. They lay waſte for thirty years, and it is only ſince the year 1780, that the Zaar Herakleus began to rebuild them; and the population increaſes conſiderably.

The plain along the left bank of the Kurr, northward as far as the river Mindoris-tſkali, (Waters of the Deſert,) is called Lillo, and only uſed for grazing of cattle. To the eaſtward it receives different names, which are borrowed from the ſtreams, marſhes, or the villages. To the N. E. ſome Aſſyrian, or Kurdic, families have ſettled, and they call it Kurdis-Mindori, Jeſidis, or Aſuris-Mindori; further E. it is ſtyled Loſchini, after ſome Oſſic families who inhabit it; and due S. twenty hours' journey from Medegh, it is denominated Niphtlik, on account of a ſpring of petroleum, from which a ſmall a ſmall neighbouring village alſo takes its name.

From Niphtlik as far as the river Kabre, or Yori, and to its junction with the Alaſan, it is called Karajoes, (Black Eye,) on account of its fine paſture. And as the ſnow ſeldom lies more than three hours, it is frequented by a large Tatarian horde, who keep their flocks there in winter. They are called Terrekemme, which only means *Nomade,* but the real name of the tribe is Serawan.

They ſtate, that they were formerly more numerous, and dwelt on the other ſide of the Araxes near Uramia; whence, ſome centuries after, they retired hither. From which, I ſhould preſume that they are deſcendants of the nation, whom Strabo mentions as living on the ſame ſpot, near the *Maſſagetes,* (perhaps Maſis-Getæ,) and calls *Saraparer.* For, the meaning, according to Strabo, of *Sar-oder* and *Saraparer,* (Beheader,) is Scythian, or Tatarian; and the *Serawan,* which in Perſian implies the ſame, underſtand that buſineſs admirably.

At the end of the plain N. E. of Teflis, only forty verſts diſtant, begins the province of Kaget, near one of the front ranges of

 Caucaſus,

Caucaſus, and is bounded to the S. by another, called Kiſik. Both formerly compoſed part of Albania; afterwards fell under the Armenian dominion; and were, at laſt, given in fief to the noble Jewiſh family of Bagarat, from which the reigning Zaars of Georgia and Iberia pretend to be deſcendants. The royal family, therefore, ſtyle themſelves *Bagration;* and have, every thing conſidered, much of the *Jew* about them.

None of the old geographers knew Kaget or Kiſik by theſe names. Moſes Choronenſis ſpeaks of Saget and Schaket, by which he probably underſtood Kiſik, unleſs ſome error in writing or printing has crept in. The Georgian Chronicle does not mention them under their modern names; for it calls the inhabitants of the lower mountainous region, *Ku-geti,* (mountaineers,) and thoſe who live nearer the main ridge, *Mu-geti,* (dwellers on the rocks.) From the former, the preſent name of the province might very well be derived; but when the other began to be called Kiſik is unknown.

Every where we ſee empty villages, and fortifications and cities in ruins; viſible proofs of a former extraordinary population.

And

And yet theſe provinces do not want men. Could they be brought from their ſcattered and inſulated caves into a larger community, they would more efficaciouſly withſtand the Leſghaes, and ſoon multiply, owing to greater ſecurity. But want of order is the cauſe; and we muſt lament, at the ſight of ſuch a deſert, that the government has not courage, inclination, and penetration, to unite the common good with their own intereſt, and to ſuppreſs the numerous body of monks and prieſts, which the ſubjection of the prince and people tends to eſtabliſh.

The road from Teflis to Kaget runs N. E. through Martcopi, an open, well inhabited, and opulent town, twenty verſts from Hallawar. The archbiſhop is always called Roſtweli. Nineteen verſts more eaſtward, runs the rapid river Yori, or Kabre, *(Caſſius Plinii,)* on whoſe left bank ſtands the village of Jaſchme. Seven verſts north lies the village of Batara-Zeul; and ten verſts E. ſtands Nino-Zminda, (the holy Nino,) a village with a convent, and a ſmall ſtone church, containing the body of the Saint, and a monument of ſome of the Georgian Zaars. The archbiſhop of this convent is always ſtyled Nino-Zmindeli. A few verſts further N. E.

 by

by E. clay huts are ſcattered on the banks of almoſt every ſtream. Further N. is the village of Sageredſcho, abounding with corn, wine, and fruits; it contains but few ſtone houſes, the reſt are mere huts.

Kudàla, Melani, and Magara, are at preſent poor villages, though formerly beautiful cities. Fine wheat is grown there, and a great deal of wine is made.

Waſchiri, a populous village, is five and thirty verſts from Nino-Zminda, and ſtands at the parting of the road. The left leads to inner Kaket over the woody mountain of Kumbet, and continues eaſtward as far as the Alaſan, a diſtance of forty verſts. On its banks ſtands Zikani, the laſt Georgian village, in ſight of the Leſghaes of Pelekan. Around its ſprings, to the northward, live ſome Duſſchic families, who are reckoned as Georgian ſubjects belonging to Kaket, and are called Schilde, Sapul, Korell, Gawar, and Makram.

To the N. W. on the right bank of the Alaſan, lies Quareli, a large village, at preſent almoſt deſerted: yet it has a ſtone church; a convent; monks; and an archbiſhop, who always ſtyles himſelf Nekreſti.

Old

Old and new Kremm, N. W. of Quareli, are well ſituated, ſtrong, but nearly half-deſerted towns, ſurrounded with indeſtructible walls, and containing a few huts, which belong to the convent; the archbiſhop of it takes the title of Kremmleli.

Allahwirdi, more S. W. of Kremm, is a fruitful and well inhabited diſtrict; and abounds with churches, archbiſhops, biſhops, and monks. Its wines are preferred to all others in Kaket. The only remarkable place is Tellaw, (city of limes,) a fortreſs impregnable to an enemy without artillery. It is in three diviſions, each ſeparated by a ſolid, high wall, and ſurrounded with a ditch, thirty feet deep and one hundred and eighteen broad. The middle and largeſt fortreſs belongs to Prince Herakleus, and is called Padonis-Zighe, (caſtle of the lords.) The ſecond, to the eaſtward, belongs to the family of Kurtſchi Bej; and the third to the ſucceſſors of Wagoſcho Swuili.

From this laſt family deſcends a Georgian monk, of the name of Cajus, who, after twelve years inſtruction in Ruſſia, returned to Georgia in 1780, and eſtabliſhed a ſeminary in Tellaw. His unwearied zeal awakened the abilities of his countrymen, and produced an infatiable defire for information. But, juft as the inftitution promifed the greateft fuccefs, that diligent and active Archimandrite was, in 1783, recalled to Ruffia with the moft flattering views. The feminary, to the great injury of Georgia, was given up, and the youths lamented their return to their former barbarifm. He remained two years in St. Peterfburg, without learning why he was recalled. In 1786, he was obliged to follow Prince Potemkin to Krement-fchug; ftill he learnt nothing of his future deftination, and remained till 1788, forgotten and almoft in mifery. In 1789, he accompanied the archbifhop of Jekatrinoflav to Jaffi. In 1792, he was again called to St. Peterfburg, and prefented to the archbifhop of Newfky. A monthly ftipend of 100 roubles was fettled on him, which the honeft monk now expends in inactivity, and pining after his native land.

Kaket, north-weft of Tellaw, is the moft populated, and beft fecured againft the inroads of the Lefghaes. The whole of this anterior range is finely wooded. Single houfes are fcattered along the banks of every ftream, and extend weftward by the river Alazohni through forefts of fruit trees.

Ahmed

Ahmed is a very remarkable ſpot; it produces good wine; the meadows are delightful, and the inhabitants very opulent.

The village of Waſchiri, above mentioned, ſtands at the parting of the roads; the left leads to inner Kaket, and the right along the mountain of Ziw. It is of conſiderable height, flat on the ſummit, and covered with wood and fine graſs. Sheep are fed there moſtly during the ſummer.

I think my conjectures are not ill-founded, when I ſtate, that Kiſik is part of *Sacaſena*, unknown even in Strabo's time. In the ancient Georgian language, this province is ſtyled Kambitſchan and Kambatſchowa. Moſes of Chorene calls it Kambidſa; and the modern Armenians, Komſchein. Strabo has juſtly denoted it *Cambyſena*. It is mountainous and devoid of water; but important with regard to the original diviſion of the countries, as the boundary between Albania, Iberia, and Armenia. But its mountainous nature does not affect the fertility in the leaſt, there being ſufficient, and, in many places, ſuperabundant, quantities of loam; which may be employed to great advantage in the fields, gardens, and vineyards, and require no watering. Yet only the hills in the

the interior are inhabited, becaufe the neighbouring Lefghaes, by their ravages, prevent all cultivation and population in the plain. They muft penetrate fecretly into inner Kifik; the ingrefs to large bands of them being impoffible, as there are but *three* ways, which are naturally ftrong, and might be effectually defended.

The *firft* is by Zminda Stepanida, (holy Stephana,) a village quite in ruins, except a handfome ftone church. Thence a barren plain, five and twenty verfts in width, extends as far as the Alafan. This plain can only be tilled in the neighbourhood of the village, as a ftrong guard would be requifite, at feed-time and harveft, to protect the hufbandmen againft the incurfions of the Lefghaes.

The *fecond* road leads, N. E. of Kumbet, through a long and hilly valley, highly cultivated and well peopled. It was formerly protected by Fort Tfchoet, which the Turks have deftroyed, and afterwards rebuilt under the name of Top-Karagatfch. At prefent, there being no Turks, it lies defolate. The view from it is delightful; we obferve with pleafure the fcattered habitations of the Lefghaes, and forget our hatred towards them, as long as the eye is feafted with the majeftic jeſtic courſe of the Alaſan. Very excellent cotton is grown in Kumbet, and the charming woods and gardens in the vicinity are well known.

The *third* road is by Tſiandghe, that opens to the ſouth of the plain, and lies between the Yori and the Alaſan. Here Kiſili, a ſtrong place, merits attention, which has the beſt land for tillage, vineyards, and gardens; and, on account of the particularly fine meadows, the breed of ſheep and horſes is very conſiderable. To the N. W. lies Anaſgis-Kari, a rugged, ſteep glen, through which one may paſs, but with great difficulty, from Karajoes to Kiſik.

In the interior of the province ſtands Padonis-Ghew, a village ſurrounded with a thick wall, containing few inhabitants. The figs there have a peculiarly pleaſant flavour. Segnak is alſo remarkable; for, as about twenty years ago, the Leſghaes made very extraordinary irruptions in Kaket and Kiſik, the Zaar Herakleus had ſome of the mountains and valleys fortified, where the people might retreat and be more ſecure; it was, therefore, called *Segnak*, (place of refuge.) But it was labour in vain; for the fortified parts remain depopulated, though belonging

to

to the Zaar. The Leſghaes have concluded a particular treaty with the inhabitants, by which the former are not allowed to ſell a man after being ſtolen from Kiſik or Kaket, if he is able to ranſom himſelf with two *tumanns;* and though the priſoner ſhould be of ever ſo high birth, yet that ſum is never exceeded. But, if a Leſghae is taken, he muſt redeem himſelf with three Georgian ſlaves; and if he be unable, his life is ſacrificed to his inadvertence, or he muſt ſeek ſome other means of liberating himſelf, without becoming burthenſome to his family or to the tribe.

Obſervations on the Quality of the Soil, Air, Productions, Population, Manner of Living, Occupations, and Superſtition of the Inhabitants of Georgia.

Kiſik wants water; and the few ſprings there are being brackiſh, can only be drank by thoſe who are accuſtomed to them. Beſides, water is not always to be had, for the villages are often ſome verſts diſtant from the ſprings; the natives are, therefore, obliged

obliged to alleviate their thirſt with fruits or wine. The ſprings are conducted with great care into brick or wooden ciſterns, from which the cattle are watered; and the over-ſhot water is reſerved for the mills. A miller places, in an inclined direction, over a wheel three feet broad, a pipe rather more than an inch thick, and about ten feet long. The water flowing from it is immediately caught for the next, and uſed in the ſame manner. But, as very little corn can be ground at a time, on account of the ſmallneſs of the mill-ſtones, the number of water-mills ſupplies the deficiency. We may reckon eighty in the valley, which is nearly fifteen miles long; and three and forty in Anaſgis-Kari. After ſupplying the mills, and the inhabitants and their cattle, the water is ſo much diminiſhed, that at the end of the valley only the traces of a ſtream can be perceived.

The air in Kiſik is purer than in Kaket. The inhabitants appear healthy and ruddy-cheeked; and the brackiſh water increaſes their appetites. The two provinces contain about 18,000 Georgian-Chriſtian families; the Mahommedans are few; and Jews are not tolerated, though they are very numer-

ous

ous on the other ſide of the Alaſan, which at preſent ſeparates Georgia from Schirvan. Its old name was *Albanus*, (Plin.) *Aluzon*, (Strabo.) The denomination Alaſan appears to be a corrupted Caucaſian word, for the Leſghaes call every river *Oeſan*, with the article *al* or *il;* whence, probably, has ariſen Alaſan.

Great part of the inhabitants have ſettled after the manner of Iberian families, and live very happy in their delightful foreſts, as long as they do not ſee a Leſghae. Agriculture and breeding of cattle are much attended to in both provinces; cotton grows in abundance; and the culture of ſilk is beſt promoted in Kaket, particularly in Lalliſkur, Lomis-Zighe, Kui-tſkarro, and Kremm.

The provinces furniſh different well-flavoured wines, but that of Kiſik is the ſtrongeſt, ſoon ſeizes the head, and requires water. In Kaget, on the contrary, they are light and very pleaſant, eſpecially that of Ahmed. But the want of caſks, and ignorance in making them, prevent their keeping above twelve months. The Georgian, however, like a determined hard-drinker, is very well ſatisfied; old wine is not to his taſte, but the new preſſed juice he drinks with great pleaſure.

ſure. The wine is preſerved in large earthen pots covered with a board, and buried in the ground; a ſmall opening is left for a reed, until the fermentation is over, when it is cloſed.

Pheaſants are very plentiful in Georgia. In Kiſik and the adjacent plains, deer and gazelles are found alſo in great abundance. On that account the chace is an eſſential pleaſure to the natives, in which they employ ſtrong, well-trained dogs, and falcons. With the latter they hunt deer and gazelles in the following manner: from the very firſt the falcon is taught to ſearch for and carry off his food from between the horns of a goat; through hunger he is rendered more and more watchful; and when he has attained to the proper age for the chace, he is obliged to fetch his food from off the head of a goat grazing at a diſtance. After acquiring this dexterity without ever failing, the day before the chace he is deprived of nouriſhment; and, as ſoon as the hunter diſcovers a ſtag or a gazelle, the hood is taken off and the animal let looſe. In the hope of finding his accuſtomed prey, the falcon darts at the forehead of the animal; which, on being caught, ſtands ſtill or falls down, and gives

gives the hunter time, either to take it alive or to kill it.

Poultry is very well flavoured; but, in general, they muſt be caught in the fields, or ſhot from the trees, as they run about wild.

Foreſts of timber and fruit trees abound in the provinces. Figs, pomegranates, apples, cherries, and cheſnuts, grow wild; and the peaches of Kiſik are much eſteemed for their particularly fine flavour.

Yet the great fertility of the country does not render the inhabitants effeminate. Under proper direction they are really very brave, even againſt the Leſghaes. But the Zaars always ſee with indifference, the Leſghaes carrying off their incautious or feeble ſubjects. Incredible as this may appear, yet it is true; for the more the inhabitants dread the Leſghaes, and are obliged to exert all their vigilance not to be kidnapped, the more the Zaars think themſelves in ſecurity, and that the miſery of their ſubjects favours the duration of their government!

Eccentricity in their piety, and ſuperſtition, are peculiar to the Georgians. A ſpoonful of water, in which an old gold medal of the emperor Conſtantine, heated red-hot, has been cooled, is an uncommon ſtrengthening draught for pregnant and lying-in women; and as they always expect to find medals at the Zaar's, he is often obliged to prepare that kind of gold-water.

The aërial ſpirits, they ſay, delight in being with the mother and her child, and muſt be very injurious to them; on that account, a dagger or ſabre is hid under the wife's pillow. Over the coverlit is drawn a red fiſhing-net, to which leaden balls are faſtened, that it may lie equally ſmooth, not ruck up, nor ſuffer any evil ſpirit to ſlip in underneath. In this ſtate the poor tormented woman is obliged to remain forty days, before ſhe dare leave her bed to breathe the freſh air. But, a female would be expoſed to the moſt dangerous ſituation poſſible, according to their prejudice, if, during her lying-in, ſhe either cut her hair or nails.

The Georgian women of rank are always attended by their parents and friends, who, with the confeſſor and phyſician, are in the next room, liſtening to her moans during labour. From them they pretend to augur a greater or leſs degree of courage in the woman, and the future deſtiny of the child, particularly ſhould it be a boy. Old practiſed wives endeavour, therefore, by the melody of their moaning, ſighing, and crying, to attract the attention of the liſteners; but, as young married women have not yet learnt that art, one of the midwives preſent imitates ſo well the moans of a lying-in woman, that the liſteners, who are ignorant of the deceit, are aſtoniſhed at the experience of the lady in the ſtraw, and loudly extol her.

Every Georgian, of the leaſt property, has a long letter, of indulgence and recommendation to St. Peter, laid on his breaſt when he is buried. On thoſe occaſions, the Monks from Jeruſalem, who pretend to have received the neceſſary credentials from the Greek patriarchs there, ſell them tolerably dear.

In Georgia and Iberia, the Holy Sepulchre poſſeſſes many famous eſtates, which have devolved to it by legacy. And, generally, men and women, when they find they muſt die, ſend their beſt ornaments, gold, and houſehold utenſils to Jeruſalem, through religious ſuperſtition.

During long continued droughts, every pariſh prieſt is accuſtomed to read public prayers, and a penitentiary proceſſion is held. Marriageable young women, of whatever

rank

rank or family, muſt then walk in pairs, and carry an ox-yoke on their ſhoulders. Between them is faſtened a cord, to keep them in their ranks, and ſerve as a clue to the prieſt who follows. All go barefooted; they wade through rivers, pools, and moraſſes; they pray, ſcream, whine, and laugh with ſo much wantonneſs, that this faſt-day has all the appearance of a licentious jubilee. The beautiful Georgian girls entirely lay aſide modeſty and good breeding; they expoſe their faces, necks, and boſoms, and frequently bring over to their wiſhes the young men, who were eager to be preſent at the ceremony.

Origin of the Name of Georgia.

Nothing is more uncertain than the hiſtory of thoſe people, who, in rude ignorance, roſe as rapid from their obſcurity, as they fell back into it. A circumſtance, though ſlight, often clears or obſcures the moſt beaten track; where the ſlave, under the laſh of his tyrant, neither recollects the paſt, nor thinks of the future.

So great is the ignorance of the wretched inhabitants of the ſouth-weſt part of Caucaſus, that, through the numerous cares for their ſelf-preſervation, they have forgotten themſelves and their origin.

On account of their conſtant diſſentions and wars, they as often change their names, as they ſubmit to new maſters; or divided into families, frequently aſſume the name of their chief, or of the almoſt inacceſſible rocks they inhabit.

Hence ariſes the difference between the various ſmall nations, on the ſummits of Caucaſus and around it.

The moſt conſiderable nation, that partly inhabits the Caucaſian chain, and partly borders on it, is the *Georgian.* Under this denomination are not only to be underſtood the inhabitants of modern Georgia, (as the province of Kartel, Kaket, Kiſik, and Armenia,) but, in a certain ſenſe, the Iberians.

The latter are not agreed reſpecting their deſcent; ſome hold themſelves to be *Aborigines,* or grand-children of Noah, by his ſon Japhet, who, as the Georgian Chronicle relates, inhabited the country immediately after the Flood. Others pretend, from oral traditions, that they were ſent from Spain to

work

work metals. Moreover, they derive the word Iberia from the Spaniſh *Emerita*, (now Merida;) and aſſert, that their proper national denomination of *Im-or-Emmeret*, has thus been preſerved, as a proof that they migrated from the Spaniſh province of that name.

The only teſt of their, not wholly improbable, pretenſions might be founded on this conſideration: the mines in the Iberian mountains were very rich; and that they were formerly worked is evidently demonſtrated. But, the preſent inhabitants no longer underſtand the art of mining, and the kings of Iberia would never enrich themſelves by ſuch means, on purpoſe to avoid being expoſed either to the jealouſy of their neighbours, or to the plundering ſpirit of their enemies. The policy of the country foregoes, therefore, through fear, the advantages that might reſult from the mines, and leaves them for ſucceeding ſovereigns.

If the evidences of the Armenian hiſtory may be truſted, Iberia, (or modern Kartel,) was formerly peopled by Jews, ſent by *Bughto-naſſr*, (Nebuchadnezzar,) to Armenia and its weſtern border; but which, at the period in queſtion, could not have been known by any name allufive to the Jews; for the hiftory only ftyles it *Wieriazik*, or the Upland of Armenia. But, the people fettled there, it calls *Jebrats;* (Jews;) and afferts, that from thence is derived the name of Iberia. However true what has been faid may be, I fhould rather deduce the name from the old Arabic word *Ibr*, (Jew,) to confirm the meaning of the Armenians and Iberians. Formerly, many large Jewifh colonies dwelt in Spain, who were expofed to fuch different deftinies, that their migration hither may not have been impoffible. Yet all this is juft as uncertain, as the period of time is when the Iberians firft conquered the Colchian, or rather Lazian, kingdom, and eftablifhed themfelves there, and when it affumed their name. Thus much, however, the Iberians and Georgians maintain as incontrovertibly true, *that their royal family defcend in a right line from David by the Virgin Mary*, and that they will reign uninterrupted to the end of the world!

But, whence the name of Georgia and Georgian is derived; whether it was given to the country by the inhabitants, or they received

received it from the country? is a queſtion that muſt remain a matter of conjecture.

The nation, who called themſelves *Georgi*, we know, bordered on the north-eaſt ſide of Caucaſus; but it is not mentioned in the hiſtory, that they croſſed the mountains, and ſettled in that part of the country which is now particularly ſtyled Georgia.

Pliny relates of the *Albani*, that they were the poſterity of Jaſon; that they aſcended Caucaſus, and ſettled along the Kurr between Iberia and Armenia. If we dare place faith in the Georgian hiſtory, theſe new conquerors did not change their family name; but after ſpreading as far as the left bank of the Kurr, and making it their frontier, the country was from them called Albania. In the pure Georgian tongue, that tract of country is called Albania, or rather *Albon*, known at preſent by the name of Kaket.

Now, as the Albanians, as well as the Armenians and Iberians, from the earlieſt times, had for their common boundary the banks of the Kurr, there remains no diſtrict thereabouts, which the *Georgi* could have inhabited; conſequently, the modern Georgians cannot have received their name from *them*. For, from the ſource of the Kurr, along both banks as far as the Arakui, Iberia extended, and was ſeparated from Albania by that river; but ridges very difficult of acceſs divided Iberia from Armenia; when, on the contrary, the Kurr parted Albania and Armenia.

Thus theſe became the boundaries of three nations, whoſe riſe and fall are related in the hiſtory. But not the leaſt notice is taken of the Georgians, though they had been known for a long term of years.

I am very well aware, that ſome derive the name from the Greek (γεωργὸς,) good huſbandman; but I ſhould commit a very great error in declaring, that the Georgians were, in general, *good huſbandmen*. They are rather the worſt of agriculturiſts; and would be expoſed to dying with hunger, notwithſtanding the fertility of the ſoil, if the Armenians did not labour for them.

However, to inveſtigate the denomination of a nation ſo well known, I think I may rely more ſecurely on the idiom and cuſtoms of the Aſiatics. Not one of the Aſiatic languages can expreſs the word *nation* ſolely, without uniting it with the *religion* profeſſed by this or that people. Thus the

Perſian

Perſian Mahommedan, if one of his ſect became an Armenian Chriſtian, would never again call the new convert a Perſian, but would ſay: "He is become an *Armenian,*" although he did not change his dwelling place. An Armenian, even in his own country, as ſoon as he profeſſes the old Greek religion of the Georgians, ceaſes to be an Armenian; that is, he is called a Georgian, (*Kurtſchi,*) not on account of the country, for he remains ſettled in Armenia, but on account of the *religion* of which he becomes a member. The heathen Oſs changes his name, and calls himſelf a *Georgian,* as ſoon as he becomes a Chriſtian. In fact, when great part of that nation were converted to the Greek church, their country received the name of *Upper Georgia,* under the protection of Prince Herakleus. The Turks, who inhabit the weſtern part of Georgia, never call themſelves Georgians, or their country Georgia. They ſtyle themſelves Muſſelmans, and their country Aghalzighe, after the capital.

The Albanians and Iberians, who, after the example of their neighbours the Armenians, became proſelytes to the Chriſtian faith, were, as may be eaſily imagined, men

who

who required ſome ſtronger inducement to the change. St. Nino, or whoever before undertook that weighty charge of conver-ſion, might eaſily conſent, that theſe people, who were then under the government of a ſole monarch, ſhould chooſe themſelves a guardian protector; (known to us by the name of *St. George;*) and that they, more-over, through love to the ſaint, as Neophytes, ſhould call themſelves *Georgians* after him.

That there is not the leaſt doubt of it, is manifeſt from this: at preſent the Albanians and Iberians do not acknowledge any na-tional diſtinction, but univerſally call each other *Georgians,* and their country *Kurt-ſchiſtan,* (Georgia,) and honour St. George as their tutelar ſaint. In Erzerum, which originally belonged to Iberia, there is, over one of the city gates, the image of St. George cut in ſtone, and from that the Turks ſtill call the gate *Kurtſchi Kapuſi,* the gate of the Georgians.

The Georgian even changes his name im-mediately on becoming a Roman Catholic, and calls himſelf *Frank,* as a proof, that he is only termed a Georgian on account of his religion. I think I have more than proba-bably demonſtrated, that the Iberians, as

well

well as Albanians, changed the name of their nation and country according to their *religion.*

The political and œconomical State of Iberia and Georgia.

Georgia reckons but few inhabitants, yet they might always conſtitute the greateſt happineſs of a prince, if he knew how to employ them to the beſt advantage, or was capable of it. In that part of Georgia, generally called *Iberia* or *Emmeretia,* there are between nineteen and twenty thouſand families. Now if the diligence of, at leaſt, from fifty to 60,000 men were united in cultivating a country ſo rich and abundant naturally, could a people be more happy than this? But both great and ſmall, as proud and contented in their dens, as if they lived in the moſt ſumptuous palaces, are, through ignorance, indifferent about the conveniences of a better ſtate of exiſtence. Sunk in ſloth, they neglect the praiſe-worthy diligence, or rather duty, of reaping the treaſures of the ſoil. Yet they are never more vigilant, never

never more unwearied, than when engaged in plunder or in war: not through obedience or love to their prince, but becauſe every one hopes to acquire wealth and poſſeſſions. This hope obliges them to be on their guard, and likewiſe valiant; they forget the dangers to which they are expoſed, on account of the gain; and prefer this ſyſtem, as the profits from it fall into their hands, to that, which by induſtry they can only expect to have rewarded after a period of ſome months. But, though the hope of war and plunder ſhould fail, owing to too many dangers, yet the ſpirit is kept alive; and, in extreme poverty, they would rather ſubmit to fortuitous circumſtances, than be laborious and live more commodiouſly. But this, though a great vice in itſelf, by no means conſtitutes the genius of the people; for then they would not feel the wretchedneſs under which they languiſh. They know very well how far they are removed from happineſs; they are too ſenſible of the penury of their ſtate, but have not the means of ameliorating it, owing to the want of a good conſtitution. At preſent, we cannot find, through the whole of Emmeretia, either city or village, that deſerves the name. In thick

foreſts,

forefts, or on green hills, each family lives feparate; and endure in this folitary ftate, with rude and favage firmnefs, the privation of thofe conveniences which they might fhare in focial life. And as the inhabitants, through want of a well-organized conftitution, become neither more civilized, nor their condition ameliorated; as, moreover, lawleffnefs renders them the moft abject flaves to their lords; fo their barbarifm increafes, and floth in poverty becomes their only confolation. But, even were induftry more encouraged, yet the boundlefs power and avarice of the great, would appropriate the fruits acquired by the labour of their fubjects. Confequently, they would remain the fame wretched beings they are at prefent; and, therefore, their indolence fets bounds to the avarice and tyranny of their mafters, whilft it reftrains their own individual happinefs. The lord alfo, equally wretched and poor with his vaffals, inftead of adopting means to improve his ftate, flies to contented pride; and though, wrapped up in a fheep-fkin, with a ftone for his pillow, he may experience reftlefs nights, yet he is always fufficiently mafter of himfelf to exclaim in a haughty tone, " that the Patriarch

triarch Jacob ſlept in the ſame manner, and that he has no reaſon to expect a better fate!"

If their ruler could once determine to try the happineſs of ſuch a government, as would reſult from the mildeſt climate, its great natural fertility, and the loyalty of his ſubjects; then he might enjoy a pleaſure in the well-being of his people, of which he is at preſent deprived. It is only by cruel commands he ſhows himſelf maſter, when he orders a head, hands, or feet to be cut off, or eyes to be put out.

A very different ſituation from that part of Georgia under the dominion of Prince Herakleus. The campaigns in India, in which he accompanied Nadir Shah, opened his eyes, and taught him to chooſe the medium between effeminacy and rude living. The conſtant occupations with the diſſolute Perſian princes and the Turkiſh baſhas, the inceſſant viſits of their ambaſſadors, obliged him to provide for emergencies, and think of means of procuring all the conveniences which his country offered, and his rank and family required. All advantages were ſeized; all poſſible methods were adopted, merely to be able to make a figure: and the reſt was left

left to be ſupplied by the merchants, who trade with Ruſſia, Perſia, and Turkey. The greateſt Feudatories formed themſelves on the model of the court. If this was ſplendid in its dreſs, table, and eſtabliſhment, they wiſhed to be no leſs ſo; and as each was aware of the neceſſary means, agriculture and breeding of cattle appeared to be the moſt advantageous to their limited views. Hence the Georgian peaſants, particularly the Armenian, redoubled their diligence to ſupport the expences of their needy maſters. But, the latter finding themſelves obliged, as their families grew up, to divide their property, and thereby expoſed to ,new wants, the peaſants were oppreſſed with freſh burdens. By theſe and other repeated extortions, they became impoveriſhed; and with a mournful countenance they now ſow for their lords, but ſcarcely earn bread for themſelves.

The inhabitants are, therefore, on an equal footing, in regard to poverty, with their maſters in regard to avarice; only that the induſtrious ſubjects of Herakleus, are more capable of meeting the exigencies of their prince, than the vaſſalage of the Emmeritans admits of. The land, alſo, which might

render

render the inhabitants all equally happy, and by its fertility equally rich, ſcarcely ſupplies enough to prevent hunger, owing to the badneſs of the government.

Another ſource of miſery is the want of juſtice, the exerciſe of it being only founded on fear, favour, or indulgence. The old princes of Georgia certainly framed laws, as the country required; but they would not learn the value of juſtice, and, at preſent, that fault is become a general cuſtom. Much as Prince *Wagtang* endeavoured to introduce the order of juſtice; yet all his labours proved abortive; he therefore wrote at the end of his code, as follows: "I have drawn up this book of laws; but in Georgia true juſtice has never been adminiſtered, and never will be." The turbulent times obliged that prince to quit Georgia, and end his days in Ruſſia. In modern times the rebellious Feudatory too often uſurps the juſtice to himſelf, which was due to another; becauſe the prince, through want of higher protection, can only exerciſe patience and indulgence towards ſuch rebels. Were even the advantages of obſerving the laws perfectly clear, yet the obſervation itſelf is expoſed to many difficulties. A general law was once

made

made for the good of the country, "not to fell the inhabitants any more as flaves." Every one faw the neceffity of fuch a law: but, through fcarcity of money, they could find no eafier way of procuring it than by entailing mifery on their fubjects, whom they fold as good Chriftians for Turkifh gold. And when thofe poor wretches became inconfolable at the feparation from their parents, from their wives and children, they were confidently advifed to be of good courage; as that was juft the way to complete their happinefs: they might become even Beys of Egypt! If complaints of this infringement reached the ears of the prince, the nobleman did not defend himfelf fo much in his right as lord over his fubjects and their property, but pleaded the example of his neighbours, who had done the fame. The tranfgreffors being too numerous, the execution of juftice was deferred, and naturally ceafed totally afterwards.

This indulgence made another nobleman, of the firft rank, fo audacious as to fell at once a whole village of Jews to the Turks; and in his defence he affirmed with great earneftnefs, like a Georgian orthodox: "In felling the Jews he had no other object,

 than

than to avenge the blood of Jeſus Chriſt!" Such an aſſertion in the cauſe of religion, and the good approving prieſts, liberated him from the puniſhment he had incurred. Georgia would, in ſhort, have become ſtill more depopulated, if the Turks, in their treaty with the prince, had not been more ſincere in agreeing not to buy any Georgian ſlaves.

The province is oppreſſed by another no leſs ſerious evil, founded in the diſunion of the nobility, and the connection of the country with the neighbouring petty Perſian princes. The Georgian nobility are too numerous, ſo as generally to poſſeſs ſufficient property to anſwer their expences. The pride of being a nobleman takes poſſeſſion of him, he reckons it diſgraceful to procure a neceſſary ſupport by the labour of his own hands, and therefore prefers indolence. In this ſtate he thinks of a thouſand artful ſchemes to plunder the property of his friend, his neighbour, and of the weakeſt. A greater or ſmaller party, equally evil-minded, unite together and mutually ſupport each other; and their falſe evidence, and the fear of their rebellious ſentiments, oblige the judge to be favourable, and give judgment for them. In a ſhort time another, ſupported by a more powerful party, contrives to gain a deciſion in his favour; but that, likewife, a frefh confpiracy annuls. Thus the fecurity and well-being of the fubject are entirely facrificed to the clafhing interefts of the nobility, and the partiality of the judge. This injury to the country proceeds, alfo, from the union with the Perfian princes, who are always at war amongft themfelves. Self-prefervation, or the defire of poffeffing the property of another, difunites them; and with a handful of robbers they declare war againft each other. Whoever, either by prefents or money, has been able to form an alliance with Georgia, receive a corps of from three to four thoufand men. And though the campaign is often ended without the lofs of a fingle man, (for both armies are more thirfty after booty than blood,) yet agriculture becomes neglected, and the inhabitant returns hungry to his empty hut, where the menials of his lord are infolently waiting to exact the prefcribed impofts.

Such is the political and œconomical ftate of Georgia: fo convenient in fituation with regard to trade, for an excellent road of only nine days' journey conveys the goods of the merchant from the Cafpian to the Black Sea. So little is the value of that happinefs known, which might be enjoyed by the introduction of commerce! But, with all this, I am aſtoniſhed that, from the continual turbulent ſtate of affairs, the revenue of the country ſhould be ſo conſiderable. I will not reckon the profits ariſing from vineyards; nor thoſe from agriculture, that pays a fifth part of the crop; but will merely eſtimate the revenue according to the inhabitants, which I reckon very much below their real number. I admit only, that Georgia, under its preſent government, is inhabited by 61,000 families, divided as follows:

Teflis, with its territory,	20,000
Thiulet and Ghefzur - -	4,000
Schemſchettil - - - -	4,000
Kartel - - - - - - -	4,000
Kaket - - - - - - -	12,000
Kiſik - - - - - - -	6,000
Karajoes - - - - - -	2,000
Somgeti and Bembek - -	6,000
Caſak - - - - - - -	3,000
Total	61,000 families.

How happy would the ſubjects think themſelves, if the fifth part of their income defrayed their taxes! They would give up a fourth, a fourth, or the half, very willingly, to ſatisfy the inordinate demands of their maſter. But I will take the eſtimate at leſs, and tax each family only ten roubles; this amounts to 610,000 r.

The cuſtoms are annually farmed at		25,000
The mine of Akdale, though worked in a moſt ignorant manner, produced, in 1780, in ſilver	- -	60,000
And in gold	- - - - - - - - - -	3,200
Erivan pays a yearly tribute of	-	15,000
	Total r.	713,200

How much might theſe revenues be increaſed, if equity and induſtry were the ſupporters of the country.

Physical Constitution of Western and Southern Caucasus.

In ſome places, near the junction of the Kubanic and Gordiæan mountains, the limeſtone becomes metalliferous; but, it has neither the direction nor the magnitude of a real a real chain of mountains. For example; near Durduſchu, in the great Avaſa, we find a milk-white, ſemi-tranſparent marble, which would be ſuperior to that of Paros, were it not filled with micaceous iron, and veins of argill. The mountain, where the famous cave of *Oggin* is, conſiſts of a dark-grey compact marble, with white and red ſpots; which, on the eaſtern ſide, is ſeen paſſing into ferruginous calcareous ſpath, variegated and luſtrous. It contains petrifactions, moſtly ammonites, with pyrites. I ſaw ſome of nineteen inches in diameter. Near the firſt village of the Abaſeks, I found an elephant's tooth in good preſervation. Similar foſſils are ſaid to be frequently diſcovered.

The Gordiæan mountains are rugged, run in contrary directions, and terminate often abruptly. Yet, there are intermediate valleys which furniſh excellent paſturage; and the foremoſt and middle ranges are covered with foreſts of oak and beech.

A reddiſh-brown *felſtein,* united with quartz, in perpendicular *ſtrata*, appears in low places and near Soghumi. It is covered with limeſtone; for, the whole weſtern ſide of the great Caucaſian range ſhows no veſtiges of argillaceous ſlate; although, in the

Kuban,

Kuban, it underlays the calcareous mountains. It ſeems as if argillaceous ſlate were produced in low moiſt ſituations. The Gordiæan mountains lead one to conjecture a very powerful agitation of the ſea, and the limeſtone indicates a ſudden and violent alluvion. For, moſt of the ranges, particularly thoſe of coarſe limeſtone, are quite of a rocky nature, not ſtratified, and never contain petrifactions. The bald ſurface of the calcareous middle-mountains bear marks ſtill of the waves having acted upon them; and, it is very evident, that the Black Sea has been much higher.

The looſe *ſtrata* dip, without exception, to the weſtward; and they prove that the greateſt impetuoſity of the waters was from W. to E. Some low calcareous hills are ſtratified, and frequently contain teſtaceous petrifactions, with impreſſions of fiſh and graſs. Others, not ſtratified, are compoſed of white calcareous earth, ſoft and fine; and, near the village of Rhaaz, the Abghazs collect very fine chalk as an article of trade.

The Avaſa abounds with iron-ſtone, but the inhabitants keep the mines entirely concealed. What they uſe in preference is of a compact texture, breaks of a ſteel colour,

 and

and is mixed with mica and calcareous ſpath. Hematite, and reniform iron ore, are alſo plentiful. Theſe ſpecies are ſmelted in a ſmall open furnace, by means of a hand-bellows. The flux is argillaceous ſand and ſcoriæ. The ſmelting is perfectly eaſy, but the ſeparation is very difficult; for, a workman requires eight hours to produce twenty pounds of iron, as his crucible holds only that quantity. The iron is ſoft and very malleable. Before the bars are uſed for other purpoſes, they are expoſed to a white heat ſeveral times, and burnt with argillaceous ſand; whereby the hardneſs is conſiderably increaſed, without injuring the malleability. The bars are manufactured into knives, ſabres, and guns, which are much eſteemed in Conſtantinople and Egypt. Yet, the natives prefer European old iron to their own; ſuch as ſabre-blades, and particularly thick wire: they are heated white and burnt at different times, to vary the degree of hardneſs, and are afterwards welded together.

Between Soghumi and Lazia, a bed of brown, compact *felſtein* runs in a north-eaſtern direction, but is not ſtratified. The low mountains, incumbent on it, are compoſed

poſed chiefly of hard ſand, or ſand mixed with calcareous earth; or, as in Mingrelia and Iberia, of indurated pure limeſtone, that extends quite to the ſea. In the plain, near the city of Odiſchi, the limeſtone is covered with black chalk, and heath-mould; but, nearer to the ſea, and at the mouth of the Phaſis, with argillaceous marl.

The banks and bed of the Phaſis, from Rhaedſch to below Cotatis, are calcareous. The fragments of rock that lie in the channel, form ſingular caſcades. Near Sannawardo, limeſtone is found incumbent on *ſtrata* of looſe ſand; and, near Kekut, the banks of the above river are compoſed of compact argillaceous ſandſtone.

Many of the lower mountains, particularly in the bay, and at the eaſtern extremity of the Black Sea, conſiſt of looſe *ſtrata*; which are become viſible in ſome places, owing to their having been furrowed by the rain and the ſtreams. I counted one and twenty different kinds, dipping regularly to the W. between 14 and 15°. The greater part conſiſt of ſand and pebbles, or looſe ſand; others contain indurated carbonate of argill, or argill with ſand and recent cockles. Between the ſtrata is often found a brown,

or

or black and yellow, iron ore, mixed with calcareous earth and mineralized cockles.

The plain of Mingrelia and Iberia, which is overflown by the Hippus at high water, was employed not long ſince for gold-waſhings; and ſome Jewiſh families had the excluſive privilege, as long as the Turks were maſters of thoſe provinces; but, when neither the Mingrelians nor Iberians would any longer allow of the works being carried on, the Jews relinquiſhed them.

The banks of the Quirila and Tſkal-Ziteli are of a ſhivery nature in the low mountains; but, in level ground they generally conſiſt of a ſoft ſpecies of ſtone, mixed with argill, ſand, and ſhells in good preſervation.

The brown *felſtein* between Soghumi and Lazia, as mentioned above, extends to the mountains of Letſchghumi and Rhaedſch, where it lies over decayed granite, interſperſed with mica and blende. When it is interſected by numerous veins of quartz, it becomes ſofter and paſſes into argillaceous ſchiſt; as it decreaſes in hardneſs, the metal appears either in veins, or in the form of a coating.

To the north of the Letſchgumi village of Zeol, in the valley that ſeparates the mountains

tains of Mingrelia and Iberia, is a large cavern of horn-ſchiſt, thickly coated with metal. And, three verſts north-eaſt from thence, whitiſh-grey rock-ſchiſt is ſeparated by an indurated argillaceous cleft, ferruginous, and containing galena in large *folia*, with ſilver and antimony.

Mining might be carried on to the greateſt advantage in the neighbourhood of Lytſchu and Kemme, as well as in Cumſi, becauſe the horn-ſtone ſchiſt conſtitutes the major part of the interior great chain. I am ſtrongly inclined to believe, that the whole chain is impregnated with it; for, I found it there under the ſame form and direction, (namely, N. W. by W. to N. E. by E.) as on the northern ſide near Bakſan and Karaghay, (ſee p. 291. vol. 1.) Moreover, it appears evident, that the whole *Porta Cumana*, on the ſouth ſide, was opened by manual labour, and the preſent large valley formed by the excavation of the mines. The ſides, and perhaps the rich mines, are now covered; yet, wherever the rock is bald, the horn-ſchiſt intervenes, and runs in a north-eaſtern direction to a great diſtance: but, to the northward, is cut off by the anterior mountains of grey *felſtein*, that contain no metal. Valleys almoſt impaſſable, between high mountains of granite, muſt be paſſed, before one can deſcend to the northern entrance of *Porta Cumana.*

If, from following old accounts, we wiſhed to determine the daily produce in gold and ſilver by the number of mines; this, including all other means of aſſiſtance, can have been effected only by a large body of men. But, as a large body in a confined ſpace muſt have impeded each other, the ſole conjecture is; "that the ore was ſo abundant, ſo rich, and eaſily obtained, that a ſmall number of men might have ſupplied the fixed daily weight of gold and ſilver." Surely this was very poſſible in a country where whole ranges of mountains are full of metal, which could have been obtained from the ſurface without much mechaniſm. To the eaſtward, half a verſt from Cumſi, ſtands a rock of grey *felſtein,* 430 feet high, and highly impregnated with galena; to all appearance, the interior muſt be equally abundant. In ſome places the mixture is of brown and grey rock-ſchiſt, covered with argill, and the galena in it incloſes much ſiliceous pyrites, under different forms of cryſtallization. But this ſpecies of mountains

tains are greatly expoſed to decay, and the metallic earth is waſhed down by rain and the melting of the ſnow; hence we find great quantities of lead-duſt on the banks of the ſtreams, and on the flats which have been covered with water after the ſnow thaws.

Near the village of Kudari, in the province of Rhaedſch, the firſt calcareous range, adjoining the principal chain, contains whitiſh-grey iron ſtone, with veins of an extraordinary ſize.

I received from Quria a piece of marble extremely beautiful: it is whitiſh-green, with large black ſpots, hard, and rather ſandy. They aſſured me, that the range of mountains from which it came, were above a verſt in length and of one ſolid maſs.

In Waghan I procured a piece of rich grey copper ore, but the miſtruſtful Awaſitſe Kaighoſro, a nobleman of the country, would not ſhow me where it was found; although I pointed out to him all the advantages which might accrue from it. He was fearful that it would excite jealouſy amongſt the princes and the Turks.

Black flint and pieces of cornelian, agate, and calcedony, abound on the road between

Cotatis

Cotatis and Waghan. They lie either between ſoft calcareous layers, or in yellowiſh red ſand; but many are ſcattered about. I found one maſs of 180lbs. weight compoſed of cornelian, agate, and calccdony; interſperſed, on one ſide, with tranſparent quartz.

I had a ſecond opportunity, in that neighbourhood, of obſerving the decay of yellow flint. It exfoliates into *laminæ*, which loſe all their tranſparency. Their interſtices are filled with the fineſt powder, that does not effervеſce with acids, but diſſolves ſlowly in nitric acid; in which the *laminæ* fall to pieces without diſſolving.

Twelve verſts from Cotatis, near the river Tſkal-Ziteli, is the famous convent of Goelette. Near it the looſe *ſtrata* of ſandy gravel contain, amongſt other fine ſiliceous ſpecies, pieces of jet. The monks call it black amber, and have the beads of their roſaries made of it. I do not recollect having met with ſuch beautiful jet elſewhere: it is very hard, and receives a high poliſh.

Twenty verſts north of the convent are low marſhy flats, covered with hills of pebbles and calcareous gravel. Some of them throw out black petroleum with ſulphureous water. The inhabitants conduct it into

deep

deep pits, for the ſake of collecting the petroleum.

The calcareous rocks of the anterior Moſſian ranges ſtretch into the ſea, and the iſlands of Sa-Cacawo are of pure ſandſtone. The former are tolerably high to the eaſtward, but the elevation is much more conſiderable at their junction with the Coraxian chain. The whole ſouth-weſtern ſide, near the Black Sea, is compoſed of ſolid coarſe limeſtone, incumbent on a browniſh-red, compact *felſtein*, vertically ſtratified. The latter extends to Akiſka, on the borders of Iberia, twenty verſts further to the S. E.

Some of the clefts in the *felſtein* are filled with carbonate of lime, or calcareous ſpath, which frequently contains yellow copper ore: being partly decompoſed, it tinges the ſpath of a beautiful green.

The Coraxian mountains are, generally, very rich in various ſpecies of copper. They alſo contain thick beds of ſand, of a fine grain, and of a greyiſh-white colour. The quarries there furniſh whet and grindſtones, which are particularly prized in Aſia Minor. But, as the ſtone in its rough ſtate is too ſoft to be ſerviceable, the quarry-men boil it in linſeed oil; whereby it acquires a greater degree

degree of hardneſs, and becomes black. Then the yellow mica appears, which was not diſcovered before.

No leſs worthy of obſervation is a cavern forty verſts from Akiſka. It is called Areſian, as well as Black Spring, *(Kara-bulak,)* from being the largeſt ſource of the Kurr. A white, ſemi-tranſparent ſteatite, called *jaſchme* in the language of the country, is prepared there, and manufactured into ſlabs, mouth-pieces for pipes, handles for knives and ſabres, cups, borders for chimneys and wells, ſhells and ornaments for baths: they are ſold at a tolerably high price. It may be the ſame cavern mentioned by Strabo in his deſcription of Galatia. I am, likewiſe, of opinion, that this mineral was carried there in his time from Akiſka; for, during my reſidence in Angora, I frequently inquired after the quarry in Galatia; but every one aſſured me, that the *jaſchme* was not produced in the country, but imported from Kurtſchiſtan.

Seven verſts eaſt of Areſian lies a beautiful marble quarry, which employs many hundred men. It is milk-white, ſemi-tranſparent, compact, fine-grained, and luſtrous. It is employed for tomb-ſtones, door-poſts, baths,

baths, and for pavement. The ftone-mafons in Akifka fell it to very great advantage.

Marfh fprings of greafy yellow water rife a few verfts from the *jachme cavern*. Where it evaporates, it depofits carbonate of foda. Every perfon carries away as much as he pleafes for foap-boiling; but it imparts a very dark green colour to the foap. Similar fprings are not uncommon in the province of Akifka, and in Oldi the water is ufed as an ingredient in the preparation of borax.

A branch of the Coraxian mountains, feparated by the Kurr, runs parallel with the river feven and forty verfts, in a northern direction, and then divides. The eaftern arm reaches the Tfanic mountains; the weftern, as foon as the Kurr winds to the eaftward, ftretches a little northwards, and unites with the middle ranges above Suram. On the latter ftands the old caftle of Awalis-Zighe, near which a copper-mine is worked.

Near the village of Zfghra-Mugha, at the head of the plain where the Kurr runs, are *ftrata* of pit-coal. There are others near Suram, on the Surimela.

The whole fouthern fide of the middle range of weftern Caucafus, from Suram,

 Xilwan,

Xilwan, and Sagina, as far as Aghalghori, merits inveſtigation, on account of the rich minerals. Some years ago, the Zaar Herakleus worked the fine mines of Sagina, but, for want of intelligent workmen, and from a too ſtrong oppoſition of the nobility, they did not ſucceed: although there were the richeſt veins of auriferous native ſilver, lead, and copper; and the neighbourhood is well ſupplied with wood and water.

Dœring, a ſkilful miner, from Eiſleben in the county of Mansfeld, worked the mines with very great ſucceſs from 1779 to 1781; but he was baſely aſſaſſinated by his comrade Dubron, of Warſaw. Conſequently they were given up after his death.

The richeſt ore was a ſulphuret of lead in aggregated ſmall *ſpiculæ,* intermixed with lead ore and calcareous ſpath on ſandſtone. A rich grey copper ore was alſo diſcovered, with iron pyrites, on a matrix of blueiſh-grey, indurated, argillaceous limeſtone, that ſoon decays on expoſure to the air.

Near Aghalghori are iron mines, and caverns filled with ferruginous earth and pieces of hematite, mixed with great quantities of ſtalactitical ſulphate of iron; ſome of the ſtalactites are five inches long, and two in thickneſs.

thicknefs. This mineral is much ufed by the dyers and tanners in Georgia.

Sulphur is alfo found in the form of whitifh-grey pyrites, between beds of yellow unctuous argill, under the mountain on which ftands the ruins of the caftle of Mtiskali, (virgin of the mountain,) near the river Xan. But it is neglected, as Georgia imports fulphur from Armenia and Turkey.

Maghal-Mindori, or the high plain, lies to the N. of Aghalghori. It is fo called owing to the junction of the fummits of the moft elevated middle mountains, which form a verdant plain feven verfts in length and three in breadth. It ferves as fummer pafturage for fheep and cattle. The northern fide is covered with fragments of bafaltic columns. Their form is the moft perfect of any that I have met with.

At the end of the glen, which has been formed by the rapid courfe of the Xan, a branch of the anterior mountains lofes itfelf, near the city of Ghori, in a mafs of trapp two and forty feet high, two hundred and twenty in length, and nearly nine hundred in breadth. The whole rock was formed into a citadel by the ancient inhabitants of Kartalizia, which they called Uplis-Zighe,

 (Caftle

(Caſtle of the Lords.) The walls, gates, ſtreets, houſes, ſhops, and even a market-place, are hewn out of the rock; and every plan has been adopted to render it impregnable, at leaſt to Aſiatic troops. But ſtill, the all-deſtroying Leſghaes, (of whom the Georgians ſay, that the graſs trampled under their feet does not ſhoot up again for ſeven years,) found means of taking it, and carried off the inhabitants.

We find many ſimilar works in Georgia, which are now become nearly inacceſſible. Not far from Duſched, in the Domain of Podorna, is a high, broad maſs of rock, where the ſtairs, even from one ſtory to the other, have been hewn out. And the area is ſo large that, in war time, old men, women, and children, flee there for refuge.

The hilly plain near the Kurr, to the ſouthward of the Domain of Ruiſch, is compoſed of ferruginous limeſtone of a laminated texture, from which flows warm ſulphureous water. Near the ſprings, the *ſtrata* conſiſt of compact, whitiſh-brown oolite, which is employed for building. It takes a good poliſh, and the church is ornamented with it.

The Domain of Mughran deſerves attention, on account of its very great fertility. Wheat, barley, and millet, are produced in incredible quantities; and a nobleman, (of the name of Mughran Padono,) aſſured me, that the two firſt never yielded leſs than thirty-fold; and that, one ſeaſon, they yielded eighty-fold. The fields are covered with blackiſh-grey vitreous lava, of which many of the pieces are tranſparent.

Urbuis is equally fertile, particularly the patriarchate of Medegh, ſituated more to eaſtward. The wine is very ſtrong and heady. The fields there, likewiſe, are covered with lava.

The mountain of Kumlis-Zighe is compoſed partly of coarſe gravel, and partly of cohering fragments; but its interior and ſummit are of brown rock-ſchiſt, vertically ſtratified. This mountain commands a moſt noble proſpect, that vies with any in Italy or in Swiſſerland. To the N. E. we ſee the habitations of the Ghefzur; to the S. W. the numerous Thiuletic vales covered with houſes; and to the N. the dwellings of the Oſſi, with the accumulated craggy points of Caucaſus. The eye follows with aſtoniſhment the different waterfalls, which claim a ſeparate deſcription on account of their volume of water and ſituation; and the number of mineral ſprings, which riſe in all directions, is truly ſurpriſing.

A range of mountains, near Gobi, compoſed of blue argillaceous ſchiſt, run in an eaſtern direction, and unite with another range compoſed of *felſtein*. A narrow vale ſeparates the former from a range of mountains of brown, compact baſalt, columnar-formed, in tetraëdrals, horizontally ſtratified, and covered with coarſe, half-rounded pieces of rock. Although maſſes are continually falling, and the mountains threaten deſtruction, yet the Oſſic family, Kilig, who live at the foot of them, are not diſpoſed to remove. (Vol. I. p. 314.).

Large heaps of broken baſaltic columns are ſcattered about in the neighbourhood of Ghuda, a deſerted village to the W. of Gobi. Some very cold ſprings riſe from under them, where travellers are accuſtomed to ſtop and refreſh themſelves, after the fatigues of climbing the mountains.

The Thiuletic mountains conſiſt of fragments of granite with whitiſh-grey or brown *felſtein*, not containing metal. Alluvions of coarſe gravel lie on them, the baſis of which is a browniſh-white, compact, fine-grained ſand-rock, that underlays the anterior mountains likewiſe. The calcareous mountains, which ſpread into a plain near Annanur, and ſtill more near Duſched, are incumbent on the above-mentioned ſand-rock. They are covered with mould mixed with vitreous lava. Although the ſoil is ſcarcely two feet deep, yet the arable and paſture lands are very fine. Vines grow amongſt the ruins of villages, and the fruit-trees offer their produce to the paſſenger. Near the Arakui large *ſtrata* of brown gypſum appear; and pure ſulphate of magneſia is found to the eaſtward of Duſched.

Throughout the whole Caucaſian mountains I had met with teſtaceous petrifactions only in argillaceous or calcareous earth. But, near Zſgetta, on the road of Teflis, there is a cleft three feet broad, in a high mountain of compact, brown *felſtein*, full of all kinds of muſcles and cockles, ſo blended together as to form a ſolid maſs, and mixed with white calcareous ſpath.

Beyond the Arakui, the *ſtrata* of gypſum ſpread out to the E. and S. and form the banks of the Kurr. But, at laſt, they are interſected by a whitiſh-grey and brown novaculite, on which Medegh and Hallawar are built.

Along the plain as far as Ghaſchme, to the eaſtward of Medegh, and to the ſouthward on to the Kurr, farinaceous gypſum is found under pyritaceous vegetable earth. In ſome places it is hardened into a laminated texture with intermediate beds of pebbles, amongſt which riſe ſprings of black petroleum. The moſt productive are thoſe near Niphtlik. The quantity collected there, in four and twenty hours, amounts to between 40 and 60lbs. The neighbouring peaſants uſe it for burning, and greaſing their cart-wheels.

The marſhes to the N. of Niphtlik, when dried up during ſummer, depoſit carbonate of ſoda; but no one makes uſe of it.

The ſcattered anterior mountains approach the middle chain near Sagaretſcho, and the gypſum there becomes metalliferous. Milk-white marble is to be met with in Kaget; but it is too fine-grained and friable, and wants luſtre. On the contrary, the alabaſter of Kremm is reckoned the beſt in Georgia.

To the W. of Sagaretſcho there is a convent and church hewn out of a mountain of brown,

brown, compact *felſtein*. Yet, this rock of ſalvation, which the monks truſted in, was ſtormed by the Leſghaes, and now ſerves for a lurking place. A plain ſkirts the foot of the mountain, and extends thirteen verſts W. The marſhes in it depoſit pure muriate of ſoda in tetraëdral cryſtals.

The plain is bounded to the W. by a chain of hills compoſed moſtly of grey aluminous ſchiſt. Springs ruſh from them, and bring down petroleum. The eaſtern ſide of the plain is compoſed of ferruginous, powdery limeſtone. Cold ſulphureous ſprings riſe there, which exhale ſtrong vapours, and appear ignited in the night-time; particularly after warm rains, and cloſe, moiſt weather. Holes have been waſhed out in particular places, and filled with flints; and they are to be found even in the fiſſures of the *felſtein*.

The inroads of the Leſghaes do not always allow the working of the alum mines. However, it is prepared in the ſame way as I have ſeen it in Danſera and Karahiſſar, in Aſia Minor. The ſtone is burnt, and expoſed to the air: after ſome days, it diſintegrates, but the alum preſerves its concretion, and is thereby diſtinguiſhed. It remains in that ſtate until its ſurface aſſumes a yel-

lowiſh-

lowiſh-red appearance, and the whole maſs becomes friable. It is then lixiviated in hot water three or four times; the lye is boiled in copper cauldrons, and poured into large ſquare boxes, where it is left to cryſtallize. Small twigs are alſo hung in the boxes for the cryſtals to adhere to.

The marſhes in the province of Kaget contain a black, ſandy, efferveſcent argill. It forms a dry cruſt in ſummer, which riſes very much in the middle, and cracks in particular parts. Its apparent ſolidity deceives; and many ignorant perſons, in attempting to croſs, have loſt their lives. Men, horſes, and carts, have been ingulphed. In ſome places there are round holes only a few feet broad; but, in others, the diameter is from ten to twenty yards; and the longeſt pole will not reach the bottom any where. This argill has likewiſe the property of undermining, and houſes and trees, which were thought to have ſtood on ſolid ground, have been ſwallowed up. Thus, for example, twenty years ago, a ſlight ſhock of an earthquake was felt, when the whole village of Waghala, with the inhabitants and their cattle, were precipitated into the abyſs, without the ſlighteſt veſtige remaining! Some of

the

the adjacent hills were rent; and ſulphureous flames burſt forth and ſcorched the neighbouring foreſts. And the people of the diſtrict were obliged to fly from the noxious vapour. Many of the trees ſtill preſent a melancholy ſpectacle of the ravages; for, particularly during the ſummer months, and in high winds, a ſulphureous vapour is emitted, and the quagmire appears in a ſtate of ebullition. And perſons pretend to have ſeen pieces of wood, waſhing-utenſils, ſkulls, and bones, and ſpoons even, floating on the ſurface. The credulous Georgians do not heſitate in declaring it to be a juſt judgment on the villagers, becauſe they were threſhing on Eaſter Sunday. Hence, being admoniſhed by ſuch an example, they heartily refrain from labour on Sundays and holy-days; and, to baniſh the remembrance of a cataſtrophe ſo truly dreadful, have recourſe to wine. A complete carouſal conſtitutes the celebration of a Georgian ſabbath, amongſt peaſants and prieſts.

The autumnal lights are as common in Kiſik as around Baku. (See Vol. I. p. 175.) And in the plain below Tſiandghe are ſprings of petroleum.

Near

Near the Kurr, and between the Yori and Alaſan, are inſulated rocks of coarſe ſand-ſtone. A convent has been hewn out of one of them, which is ſtill inhabited, and called Sa-Utawno, (The Hermitage.) The cells, church, court-yard, refectory, and cellar, have been formed by excavation; and a ſpring having been diſcovered in the church, ſerves for the baptiſmal font. With the water the holy wafer is boiled; children of the great, and thoſe who wiſh to devote themſelves to the cloiſter, are baptized; and it is given to the faithful in all kinds of diſorders.

The plain of Kara-joes, on account of the intenſe heat, the number of poiſonous ſerpents, and the marſhes formed by the ſpring-waters, is abandoned in ſummer. But the Serawan tribe, and other hordes, encamp there in winter, for the ſake of the fine paſturage. The thoughtleſs Georgians never think of laying in a ſtock of hay for the winter, although it is ſometimes very ſevere. In 1780, the ſnow lay ten days, and many thouſand head of cattle periſhed with hunger. In the plain is found a particular ſpecies of ſnipes. Owing to their ſhort wings, they fly low and heavy, and are often taken by

hand.

hand. They are highly flavoured, and are called Iſraelitiſh ſnipes.

The marſhes contain carbonate of ſoda. In fact, the province of Georgia appears to be ſo full of all kinds of ſalts, that they may occaſion the decay of the mountains, which decreaſe in height very evidently. Thoſe compoſed of coarſe granite and argill. are moſt affected. The calcareous coating of the mountains in Kiſik appears, likewiſe, to have been cauſed by decompoſition; and may it not be conſidered as the origin of the extraordinary fertility of the province?

The ſaline quality of the ſoil is, probably, the reaſon why ſo few ancient medals are dug up. An old tomb was opened whilſt I was in Teflis, which contained an iron ſhield, ſword, and battle-axe; but, they were ſo much corroded, that they fell in pieces. There was alſo an inſcription on a ſtone at the foot of it, but too much decayed to be legible; however, we could diſcover, that it was in Arabic. Near the ſtone ſtood an earthen pot filled with ſilver medals, which crumbled between our fingers. There were three hundred, ſimilar to thoſe attributed by Count de Guebelin to Choſroes, king of Parthia; but he is miſtaken.

Georgia is entirely destitute of salt-refineries. Rock-salt is generally used, which is imported from Bajazit, in Persian Armenia.

Topography of the Countries between the Rivers Terek and Kurr.

The tract of country along the western shore of the Caspian sea, between the Terek and Kurr, extends nearly 600 versts in length, between 39 and 40° N. lat. The former divides it from the government of Astrachan, and the latter from the desert of Moghan; and towards the mountains from the territories of the Khans of Schuschi and Ganscha. Its whole N. E. frontier towards Georgia is formed by a chain of mountains, connected with the Caucasian Alps, whose inhabitants are known by the common name of Lesghaes.

The Lesghaes are reckoned at thirty different nations, being less distinguished by their custom and manner of living, than by their languages. Each nation is subdivided into tribes, who bear proper names. The denomination *Tawlinzi*, common to all, only means

means inhabitant. The word Kumuk is often uſed in a ſimilar extenſive ſenſe; but, at preſent, is confined to the inhabitants of that province.

Their country is called Leſgiſtan. Its breadth varies, and is moſtly inconſiderable in compariſon with the length, if we except ſome places in the province of Schirvan, particularly in the mountains, where the Kurr conſtitutes its boundary; for there the breadth exceeds a third part of the length. According to conjectural calculation, its ſuperficies may contain about 45,000 ſquare verſts. But I cannot anſwer for the preciſion of this admeaſurement, having had no correct map of the country. The engineers are projecting one, which will probably be completed very ſhortly.

It is divided into the three provinces of Kumuk, Dagheſtan, and Schirvan. The firſt has long been more cloſely connected with Ruſſia, than with the new Perſian monarchy; whilſt the latter are either under the ſovereignty of Perſia, or, at leaſt, claimed by its regents.

THE PROVINCE OF KUMUK,

Lies between the rivers Terek and Koiſu, and is eighty verſts in length. Its diameter, from the ſea-ſhore to the mountains, may be reckoned at ſixty. The principal part is a plain between the ſea and the foot of the mountains; yet the mountains neareſt to the plain equally belong to it. The more diſtant and elevated range is inhabited by Leſghaes. It is watered by only two ſmall rivers, the Akſai and Kaſma. At their ſources the mountains are covered with foreſts, which alſo clothe the banks of the Koiſu; and a branch of it is called the *dry* (Kuru) Koiſu, being filled only when the ſnows melt. The two principal mouths are Sulak and Agrachan; the former is applied by the Ruſſians to denominate the whole river. It is between 40 and 50 fathoms broad, and contains an uncommonly rich fiſhery. The ſoil is particularly fertile.

Different beys govern this province, of whom two are the moſt reſpectable and powerful. One reſides in Akſai, and the other in Endrie, called by the Ruſſians, Andreewka. Both cities ſtand at the foot of the

the mountains; the former on a rivulet of the ſame name, and the latter on the Kaſma. They are inhabited by Armenian and Georgian merchants. Villages are moſtly ſituated in the mountains; but there are three on the Koiſu; namely, Temir, Koſtek, and Kaſiurt. The plain is peopled by Nogay Tatars, who feed their numerous herds near the canals, ſtreams, and wells. The Leſghaes alſo bring down their cattle in winter, till the ſnow on the mountains has melted. For this permiſſion they, as well as the Tatars, pay a duty to the Beys.

DAGHESTAN.

This province extends 200 verſts along the Caſpian, from the Koiſu to the little river Rubas. The breadth varies, and in ſome parts amounts to between fifty and ſixty verſts. Dagheſtan, in the vernacular tongue, means a mountainous country. In fact, the province is moſtly of that nature; for the breadth of the plain is conſiderable only towards the frontiers of Kumuk, and is ſoon ſhut in by the chain that approaches the ſea.

Dagheſtan conſiſts of the following ſmall ſtates: The territory of the Schamgal; that of the Uzmey; the Khanſhip of Derbend; and the Domain of Tabaſſeran. Their neighbours, who inhabit mountains which are never free from ſnow, are the Awarev, the Akuſchinzs, the Kubactſchi, and ſome other ſmaller Leſghaean tribes.

TERRITORY OF THE SCHAMGAL.

This title is borne by the regent of a country extending 100 verſts in length, from the Kuru-Koiſu to the ſtream Uruſai-Bulak. Its diameter, from the banks of the river to the high mountains of Leſgiſtan, may be taken at from fifty to ſixty verſts. The frontier on this ſide, ſtretches a great length along the Koiſu. On the borders of Kumuk, above the river, the mountains are about fifty verſts from the ſea; but, further on, they approach much nearer, and at Tarki, the capital, are only ten verſts diſtant. From Tarki

Tarki to the Urufai-Bulak, are reckoned fifty verfts. Here they are never more than from ten to twelve verfts from the fea, and often confiderably nearer. Corn is moftly grown in the plain. It is fertile, and well watered by the rivers and ftreams; and, where neceffary, interfected by canals. The mountains are univerfally fteep, feparated by numerous narrow valleys, and generally covered with forefts and underwood. The inhabitants live moftly on the mountains, as there are only fheds in the plain, where the Lefghaes pafture their cattle in winter and fpring, for which they pay a duty to the Schamghal.

Between the Koifu and the Urufai-Bulak, the country is croffed by two rivers; namely, the Ofeni and the Manaffa. Both rife in the higheft of the Lefghaean mountains, fixty verfts from the fea, and are uncommonly rapid; particularly in the month of May, when fwollen by the melting of the fnow. Between, an arm of the Koifu and a falt ftream called Bakas, form, near the fea, large moraffes covered with rufhes. Between the Ofeni and Manaffa ftands Tarki, on a declivity facing the plain, in a narrow glen abounding with fprings and ftreams, fome of which are falt. It is faid to contain

 10,000

10,000 inhabitants, among whom are many Armenian and Georgian merchants. It is the reſidence of the Schamghal. The reigning one, called Mahommed, is an old man of eighty, and devoted to drinking.

Between the Manaſſa and the Uruſai Bulak, are other ſtreams; on one of them, called Buinacki, ſtands a city of the ſame name, but not ſo reſpectable as Tarki. It ſerves as a reſidence for the eldeſt ſon of the Schamghal, who has a ſmall Domain attached to it.

TERRITORY OF THE UZMEY.

The regent of the country between the Uruſai-Bulak and the ſmall river Darbach, bears this title. His territory extends along the Caſpian about ſixty verſts; the breadth is nearly the ſame. It contains very little plain, as the mountains are ſeldom more than from ten to fifteen verſts diſtant from the coaſt; except near the Darbach, where they are about twenty.

Three rather conſiderable rivers, the Chamraſeni, the great Buam, and the great Darbach,

bach, which riſe in the higheſt Leſghaean mountains between ſixty and ſeventy verſts from the ſea, flow in almoſt parallel courſes through low countries, partly marſhy and woody. They are all banked up at the foot of the mountains, to ſerve for irrigating the plain. Between the Uruſai-Bulak and the Chamraſeni, the Intſche flows through rugged banks into the ſea; and between the Chamraſeni and the great Buam is the little ſtream Baſchli. Between the latter and the great Buam are other inconſiderable ſtreams and ſalt lakes, and a warm ſulphureous ſpring very near the ſhore. The Scheriff, an arm of the great Buam, flows between that and the great Darbach. A great deal of wood grows near the mouths of all theſe rivers and ſtreams, and the intermediate fields are in tillage.

The whole territory of the Uzmey is very populous. The town of Baſchli, on the ſtream of the ſame name, about thirty verſts up in the mountains, is his reſidence. On a branch of the Intſche ſtands the city of Ottemiſch, and there are many villages in the mountains. Higher up the Darbach live the Karakaideks, and many villages of the Kaideks lie on the Buam. They were originally free Lesghaeans; but, at present, are under the dominion of the Uzmey, who, on that account, is commonly styled Khan of the Kaideks. The Berekoezs, equally his subjects, live between the mouth of the great Buam and the Darbach. Besides their chief village Berekoee, on the Buam, they have others in the interior.

KHANSHIP OF DERBEND.

Its circumference is very small, and it owes its importance solely to the situation of the capital. It extends along the Caspian little more than thirty versts, between the Darbach and the Rubas; and the breadth, in-land, scarcely amounts to from ten to fifteen, being bounded by the Domains of Tabasseran belonging to Daghestan. Between the Darbach and the city of Derbend lies a marshy plain, thirteen versts in width, intersected by many small streams, and here and there converted into fine corn-fields. On the other side of the city, towards the Rubas, is a similar plain, equally cultivated. Near Derbend, the mountains stretch out towards

towards the ſhore, and are very rugged, and covered with buſhes within two verſts and a half of it.

The city is built on an acclivity, and forms a parallelogram of twenty-ſeven verſts in length, and 200 (German) fathoms in breadth. The walls, on both ſides, are continued on to the ſea; ſo that there is a ſpace of one and a half verſts long and 200 fathoms broad, ſeparated from the city by a croſs-wall called Diebari. The ſhore riſes gradually, ſo that, except boats, no veſſel can approach nearer than a verſt. The walls are built of hewn ſtones; not leſs than five fathoms high, in many places ten feet thick, and fortified with round and ſquare towers. The ramparts are narrow, conſequently only the towers can carry cannon. The citadel occupies the higheſt point, which is ſeparated from another height that commands the city, by a very narrow, deep dell, whoſe ſides are almoſt perpendicular. This citadel is triangular, and forms an angle with the city; its walls are above ſix fathoms high, and the towers about eight. The Khan reſides there. Notwithſtanding the numerous ſieges the city has ſtood, yet the fortifications are in tolerable repair. I meaſured many of the ſtones, and always found them ſix feet every way. The neighbouring mountains ſupply Derbend with fine water, and the citadel is furniſhed with a reſervoir. Gmelin's deſcription I have found correct throughout, ſo that I have nothing particular to add. According to the lateſt obſervations, the N. latitude of Derbend is 41°. 52′. 0″.; the variation of the needle, on the 16th of June, 1796, was 11°. 41′. 20″ E.

The remains of the wall carried over the mountains, as mentioned by Gmelin, ſtill exiſt: and the inhabitants, as well as the mountaineers, are all agreed that it formerly extended from the Caſpian to the Euxine ſeas, and aſſert that there are ſtill partial traces of it in the interior.

Feth Ali Khan, who had uſurped the ſovereignty of Derbend, died about ſeven years ago; and left it, with other poſſeſſions which he had partly inherited and partly ſeized, to his eldeſt ſon Achmed. Since the death of the latter, they have deſcended to his brother Tſchich Ali, who is ſcarcely eighteen, and who delivered up the city to the Ruſſians in the preceding year.

TABAS-

TABASSERAN.

This province lies between the Darbach and Rubas, nearer to their ſources; and reaches to the higheſt Leſghacan mountains, about forty verſts in extent. Thoſe mountains are particularly rugged, and covered with wood. Many villages ſcattered throughout the beautiful and fertile valleys. The government is divided amongſt five brothers, each poſſeſſing a certain number of villages. Ruſtom, the eldeſt, has by far the greateſt ſhare, and bears the title of Kadi of Tabaſſeran.

SCHIRVAN.

This province is, without compariſon, the largeſt and moſt important diviſion of the country we are treating of. It therefore merits, in every point of view, a more accurate and circumſtantial deſcription. It extends from the Rubas to the mouth of the Kurr, along the Caſpian, above 300 verſts in length. The former divides it from Dagheſtan, and the latter from the Moghanic plain,

plain, from the Khanſhips of Schuſchi and Ganſcha, and from the north-eaſtern parts of Georgia. On the other ſide it is bounded by the Leſghaean mountains. The breadth, reckoning from the ſea-coaſt, varies very much. On the Rubas it is ſcarcely twenty-five verſts; near Kuba it amounts to about ſixty; and from the mouth of the Ata nearly eighty. Further on it increaſes, and the largeſt part is from the point of the peninſula of Abſcharon, neareſt Baku, to the Kurr; where this river is croſſed by the high road from New Schamachi to Ganſcha. Between thoſe two points it is reckoned at 240 verſts. From thence it gradually decreaſes down to the mouth of the Kurr.

According to its natural properties, it may be divided into the four following diſtricts: the plain at the foot of the mountains, between the Rubas and Ata; the barren mountainous region, from the Atatſchai to the plain on the left bank of the Kurr; the plain near the Kurr; and that part of the higher chain which bounds the above three diviſions, and belongs to Schirvan.

Its political partitions conſiſts of: the Khanſhip of Kuba; the tract of country beyond the Atatſchai, conquered by the Khan of of Kuba; the Sallianic territory; the Khanſhip of Baku; that of Schamachi; and the Khanſhip of Scheki.

NATURAL DIVISION OF SCHIRVAN.

Plain between the Kubas and Atatchai.

The mountains that extend towards the ſea near Derbend, gradually recede behind the city; and approach again, in a circular form, within four verſts of the ſea at the mouth of the Ata, about 130 verſts diſtant. The greateſt diameter of the circle, between the Samur and Keſartſchai is about fifty verſts. Hence is formed a plain, between the Rubas and Ata, that conſtitutes a conſiderable portion of the province. It is irrigated by numerous ſtreams from the mountains, which are buried in eternal ſnows. Theſe certainly contribute greatly to its fertility, but render the paſſage of an army extremely difficult. They branch off in various directions, are uncommonly rapid, and, being ſhallow, have a wide channel filled with large fragments of the rocks. This is

a natural

a natural cauſe of their velocity; and as the mountains conſiſt moſtly of looſe argillaceous *ſtrata*, the water, at particular ſeaſons, is ſo impregnated with earthy particles, as to be ſcarcely drinkable. Even by boiling it the evil is not entirely removed, and filtration does not always ſucceed; the beſt method is to cleanſe it with coal-duſt. However, through want of ſpring-water, we were obliged to put up with it; fortunately, even in its unpurified ſtate, it does not ſenſibly affect the conſtitution.

A conſiderable part of the plain is covered with ſtones and earth brought down by thoſe rivers, and perhaps it owes its formation, at leaſt its extenſion, to them. It is covered with ſmall woods and clumps of buſhes; and the villages are ſurrounded with orchards, vineyards, and plantations of mulberries, and are a moſt convincing proof how diſpoſed nature is to reward diligence and induſtry.

The diſtance of the following rivers are given according to the direction of the road from Derbend to Baku. The moſt remarkable are:

The Gurgeni, ſixteen verſts from the Rubas.

The

The Samur, eight verſts from the former, is the moſt conſiderable of all. Though the depth be ſeldom more than three feet, yet its channel, being formed by many collateral branches, is nearly a verſt in breadth. It is filled with heaps of ſtones and iſlets covered with earth. In ſummer, when ſwollen by the ſnows, the paſſage is very difficult and often dangerous. At that ſeaſon it reſembles the torrents in Swiſſerland, and the impreſſion made on the traveller who croſſes it near the mouth, is greater than thoſe make, becauſe he ſees it pouring down the plain, but can ſcarcely diſcover the mountains from whence it ruſhes, owing to their diſtance. It is alſo worthy of notice, that one often perceives, at a tolerable diſtance from its bed, on more elevated ſpots, holes ſuddenly filled with water, where no appearance of it exiſted a few minutes before: a circumſtance that occurred in my preſence, at the diſtance of half a verſt from the channel. During the day, the quantity of water increaſes, and in the night diminiſhes; which evidently ariſes from the effect of the ſun whilſt above the horizon. Eight or ten verſts from the Samur, we meet with another arm of it called Jaloma. On this arm

Reineggs has erroneoufly placed the city of Kuba, as well as the fmall harbour of Nizabad at its mouth.

Fifteen verfts further on, we difcover a number of ftreams, at different diftances from each other, which include an extent of thirteen verfts. They are the collective arms of the Kefartfchai, whofe proper bed lies beyond them; and no torrent is divided into fo many.

From the main branch of the Kefartfchai to the Deli, are four verfts; on whofe right bank, where it quits the mountains, ftands Kuba, the capital of that part of Schirvan. According to an obfervation made fix verfts E. N. E. of the city, the N. lat. was found to be 41°. 24′. 57″. And at the mouth of the Deli ftands the village of Nizabad, with a fmall harbour, but infecure.

From the Deli to the Aktfchai are five verfts;

And from there to the Karatfchai five more;

Thence to the Dfchagidfchich four verfts;

From there to the Belbch ten verfts; and ten more to the Schabran. This laft is certainly not more confiderable than the abovementioned, but remarkable becaufe the ancient

cient city of the ſame name ſtood on its left bank, of which the ruins are ſtill viſible. Here it ſhould be obſerved, that the country beyond Schabran, as well as the plain along the coaſt, and the neighbouring heights, reſemble the barren tract between the Atatſchai and the plain on the Kurr; that is to ſay, they contain leſs wood and water, and are generally leſs fertile.

Nine verſts beyond the Schabran flows the Ewitſche; and

Twenty-two verſts further, the Gueljen. From its left bank, where it iſſues from a narrow valley, extends a kind of rampart, in a diagonal line acroſs the plain, to the ſea-coaſt about eight verſts diſtant. Its poſition and appearance would lead one to ſuppoſe, that it was a fortification, and there are perſons who attribute it to Alexander the Great; for, it is a common cuſtom to aſcribe to him all works of human induſtry, of which the origin is forgotten. But, a cloſer inveſtigation proves, that it has been a conduit for the waters of the Gueljin to irrigate that part of the plain deſtined to the culture of rice. On the neighbouring mountains are the remains of an old caſtle, called Tſche-

crecii,

creech, which the inhabitants equally attribute to Alexander.

From the Gueljin to the Ata, where the plain terminates, are other twelve verſts.

The barren mountainous Region from the Atatschai to the Plain on the Kurr.

The ſecond diviſion of Schirvan, marked by its natural properties, is formed by that tract of country which extends on one ſide, from the Ata, along the coaſt, to the plain watered by the Kurr; but, on the other, is bounded by the higher range of mountains, which run in an equal courſe through Schirvan. The length is about 120 verſts; its breadth differs conſiderably in many places, becauſe the coaſt projects very much between the mouths of the Atatſchai and Kurr, and forms a kind of ſalient angle between thoſe two points. Thus, from the mouth of the Suguite to the heights near New Schamachi, we may reckon about 100 verſts in breadth. The plain is never of any conſiderable breadth near the coaſt. It does not border immediately on the elevated woody region, but is ſurrounded

ſurrounded by lower ranges and valleys, barren and devoid of water, which conſtitute the principal portion of this diſtrict.

The part neareſt the ſea is the moſt ſterile; but, towards the higher ranges, fertility increaſes. Baku ſtands in a moſt dreary ſituation, where the ſoil is impregnated with ſalt and bitumen; and calcareous rocks are partially ſcattered. But the ſale of naphtha, petroleum, and ſalt, is a ſource of great wealth to the city. The want of water proceeds partly from the ſaline and bituminous qualities of the ſoil, and partly becauſe the major part of the ſtreams, which riſe in the ſurrounding mountains, do not flow directly towards to the ſea, but fall into the Kurr.

On the road from the Atatſchai to Baku, the firſt remarkable object is Mount Peſchparmak. It ſtands about ſix verſts from the river, and conſiſts properly of accumulated rugged maſſes. The largeſt, when viewed from the Caravanſarai there, appears to project very boldly; and that, probably, is the reaſon why European travellers always mention it as a remarkable object. Gmelin deſcribes it very circumſtantially. It appears to me to be rather a welcome ſight to mariners, than a ſingularity in itſelf. Behind lie the ruins of a caſtle, inhabited by herdſmen, who paſture their cattle in the neighbourhood.

From the mountain to the little river Sugaite, or Koſutſchai, we have to croſs a tract of deſert for more than fifty verſts, and from thence to Baku are thirty more. This city ſtands on the peninſula of Abſcharon, which contains not only the moſt and the beſt ſprings of petroleum and naphtha, as well as ſalt lakes, but where alſo fire may be found on every part of the ſurface. I ſhall deſcribe this ſingularity more minutely hereafter. About ſixty verſts beyond Baku, the river Pirſagat is the only one that diſcharges itſelf into the ſea; all the others, which flow from this ſide of the mountains, fall into the Kurr.

PLAIN ON THE LEFT OF THE KURR.

In the direction of the river, this plain is about 200 verſts in length. Its greateſt breadth may be between ſixty and ſeventy. It is moſtly ſurrounded by mountains. Near the banks it is expoſed to inundations, and overgrown with ruſhes to a conſiderable diſtance. The part next the ſea-coaſt is ſa-

line,

line, and perfectly ſterile; but the more elevated tract is tolerably fertile. About 100 verſts from its mouth, the Kurr receives the Aras; and on its left bank, or on the ſide of Schirvan, ſtands the large village of Dſchawat, belonging to a Bey under the protection of the Khan of Schamachi. After its junction with the Aras, the Kurr becomes navigable, and is above 140 yards broad; the banks are covered with wood. Thirty verſts from the ſea, the river divides into numerous branches, of which the outermoſt are the largeſt. The intermediate iſlands belong to Schirvan. On the left main-branch ſtands the city of Sallian; which, properly, is only a collection of villages, and owe their wealth to an uncommonly rich fiſhery. New Schamachi lies alſo in the plain, about fifty verſts from the river and equidiſtant from the ſea, on the little river Akſu, that falls into the Kurr, five verſts from where it iſſues from a neighbouring valley.

THE UPPER CHAIN OF MOUNTAINS.

I have already obſerved, that a conſiderable part of the mountains, which ſurround the three diſtricts above-deſcribed, belong to the province of Schirvan; but this is the moſt important of them all. Near the Rubas, as far as Leſgiſtan, it is very narrow; near Kuba it becomes wider; on the other ſide of the Atatſchai, by the Khanſhip of Schamachi, it gradually increaſes; and its greateſt width is in the mountains, where the Kurr forms the boundary of the Khanſhip of Scheki.

The Leſghaes, who inhabit the Schirvanic frontier, are almoſt all, directly or indirectly, under certain regulations, governed by two chiefs: namely, Uma Chan, whom the Avars acknowledge; and Khanbutai, who rules over the numerous tribe of *free* (Kaſi) Kumuks, who live in the mountains between the Samur and Deli. The two higheſt ridges covered with eternal ſnow, are: the Schachday, behind Kuba, in the territory of Khanbutai; and the Khalader, behind Old Schamachi, in the territory of Uma Chan. The latter dips very much from the parallel direction with the ſea-coaſt, and inclines moſtly towards the Kurr; hence ſo many rivulets riſing there, inſtead of running towards the ſea, fall into that river in an almoſt oppoſite direction.

I muſt juſt obſerve, that the denominations Uma Chan and Khanbutai are to be conſidered, like Schamghal and Uzmey, not as names, but as titles; thus, the reigning Khanbutai's name is Suchai-Chan.

The range of mountains, belonging to Schirvan, are interſected by many narrow valleys, which ſerve as channels for the rivulets that flow from the uppermoſt ridge; but ſprings and ſtreams riſe alſo in that range. And I have remarked ſmall lakes, in very elevated ſpots, full of water even in the hotteſt ſummers. In other places, water is found by digging only a few feet. Foreſts, coppice, and open plats, in all the pomp of verdure, embelliſh that fruitful country in a moſt enchanting manner; and it is not ſurpriſing, that the inhabitants prefer it as a reſidence.

Part of the month of June, and almoſt the whole of July, 1796, I paſſed in the mountains, between the ſources of the Ataſchai and Segaite; the place is called Kurt-Bulak, and lies in 41°. 9'. 47". N. lat.

The mountains round Kuba appeared the moſt fertile and pictureſque. Behind Old Schamachi they are, likewiſe, fertile; and near the Armenian villages there, the ſouthern ſide is planted with vines.

Old Schamachi ſtands on an acclivity where the upper ridge loſes itſelf in the barren region; about four verſts from the Pirſagat, and nearly thirty from New Schamachi. About 100 Mahommedan families now live ſcattered through the ruins of that once famous capital.

POLITICAL DIVISION OF SCHIRVAN.

KHANSHIP OF KUBA.

This Domain includes that part of Schirvan lying between the Rubas and Ataſchai, the tract of mountains as far as Leſgiſtan, and the above-deſcribed fruitful plain. It is inconteſtibly the beſt and moſt populous diviſion of the province.

The capital is ſmall, and, according to the cuſtom of the country, fortified with towers. On the ſide of the river it is open, as the high, ſteep banks ſupply the place of works. Every thing elſe relating to it has been deſcribed by Gmelin.

The

The Domain is a family poſſeſſion of Tſchich Ali, Khan of Derbend, whoſe father, Feth Ali, conquered that city, and almoſt the whole of Schirvan. At the end of laſt year, after the flight of Tſchich Ali, his younger brother Haſſan, a minor, was made Khan, under the protection of Ruſſia.

Conquests of the Khan of Kuba and Derbend beyond the Atatschai.

Before the time of Feth Ali, the territories of the Khans of Baku and Schamachi commenced from the right bank of the Atatſchai. The former poſſeſſed a tract of country on the coaſt, of a certain breadth inland; and the latter the remainder from thence to the frontiers of Leſgiſtan. But Feth Ali ſeized the major part of the Khan of Baku's, and confined him to the city with a ſmall diſtrict; and thus it has hitherto remained. In fact, this conqueſt is not particularly important, for though it is a great extent of country, yet it is the moſt barren, and without thoſe advantages which accrue to the Khan of Baku from his portion. Its population is confined to a few villages.

DISTRICT OF SALLIAN.

The city, with the ſurrounding part of the plain, lies on the left bank of the Kurr. It is a heritage belonging to Feth Ali's anceſtors, which deſcended to his ſon Tſchich Ali; and is governed by a *Saib*, or Stadtholder. The population is conſiderable, and the inhabitants are opulent. The fiſhery in the Kurr is moſtly carried on by Ruſſian ſubjects, from the government of Aſtrachan; hence, many foreigners reſide in Sallian. The Khan's revenue, for it is farmed, is ſaid to amount annually to 50,000 roubles in ſilver. Before the time of Feth Ali, the Sallianic Domain was not united with his other poſſeſſions in Schirvan; but, at preſent, by means of the conquered diſtrict, they are conſolidated.

KHANSHIP OF BAKU.

The ſmall portion remaining to the Khan, is confined nearly to the peninſula of Abſcharon, on the right ſide of the Sugaite. Although a moſt barren ſpot, it has the advantage of numerous ſprings of petroleum and

and ſalt; and the city is the only ſpacious and convenient ſea-port on the Caſpian, for a long diſtance, where inland or foreign produce can be ſhipped. It ſtands on a declivity facing the ſea, is ſmall, and ſurrounded with towers and a double wall. The remains of a dry ditch, with ſome baſtions, made in the time of Peter the Great, ſtill exiſt. A more circumſtantial account is to be found in Gmelin. The trade has very much declined, owing to the conſtant wars in Schirvan and the adjacent countries, but more particularly from the fall of Schamachi.

Salt is moſtly conveyed by land to Schamachi and other parts. The greateſt conſumption of naphtha is in Gilan, as the inhabitants of that province, who are ſilk-weavers, either taught by experience, or from a general prejudice, believe that no other ſubſtance can be uſed for burning, by thoſe who breed ſilk-worms, but naphtha, without injury to the animals. It has been already obſerved, that the peninſula contains an inexhauſtible ſtore; on that account it is the moſt populous part of the diſtrict. We might even ſay, that the total population is confined to the city and thirty villages.

KHANSHIP OF SCHAMACHI.

To this territory belongs the whole tract of country beyond the Atatſchai, neareſt the mountains, as far as the Khanſhip of Scheki and the borders of Leſgiſtan. It includes, likewiſe, part of the plain on the left bank of the Kurr. Since Nadir Shah down to modern times, it has ſuffered more from perpetual commotions than any other part of the province; conſequently, population and property have been ſeriouſly affected. Formerly, the Khan was the moſt powerful regent in Schirvan, and his capital famed for its ſize and the riches of its inhabitants; whoſe prime ſource of wealth was the ſilk-trade with Perſia: but, at preſent, he is far inferior to the Khan of Derbend; and of old Schamachi we ſee nothing but extenſive ruins. In moſt places, the walls are in tolerably good preſervation, although they incloſe a ſpace, on a declivity, of ſome verſts in circumference. The ancient city muſt, therefore, have been much more extenſive. The architecture and manner of fortifying are ſimilar to what we meet with at Derbend, Kuba, and Baku; but there appear to have

have been more maſſive public buildings in Old Schamachi. When we conſider the beautiful moſks of hewn ſtone, particularly the Bazars, in the ſame taſte and of ſimilar materials, which have ſuffered little from the ravages of time and of the deſtroyers, we are aſtoniſhed how Gmelin, who found the city inhabited, could ſay, that all the buildings were of rough materials, and no public ones claimed the leaſt attention.

The hiſtory of the Khans and of their capital, from Nadir Shah down to the year 1772, are to be found in his works. At that period, Schamachi was among the conqueſts of Feth Ali, and the family of the Khan were partly in priſon, and partly expulſed. It was again inhabited, and New Schamachi abandoned. Two years after Gmelin's abſence, as the very diminutive number of inhabitants did not bear the leaſt proportion to the circumference of the city, it could not be defended againſt hoſtile inroads, which were every moment to be dreaded from the neighbouring Leſghaes. Feth Ali, therefore, determined to abandon it again, and reſtore New Schamachi, more ſuitable to the population, and ſo it has remained.

The

The form is that of a quadrangle, each ſide being 800 paces in length. The walls are built of unburnt bricks, fortified with ſquare and round towers at particular diſtances, and a very broad, deep, dry ditch. In 1795, the houſes were deſtroyed by an irruption of the famous uſurper Aga Mahommed Chan, and ſtill lie in ruins. The number of inhabitants at that time I cannot ſtate upon a certainty, but it does not appear that they could have exceeded 6000. There are many Armenian merchants, who trade in ſilk of an inferior quality.

Since the death of Feth Ali, the former reigning family ſucceeded in regaining the poſſeſſion of the city, and the whole territory; to which the minority of Feth Ali's ſons greatly contributed. Kaſſem, ſon of Agaſſe Chan, brother of the expulſed regent, firſt made himſelf maſter of Schamachi; but was ſoon depoſed by his brother Muſtapha, who reigned till the end of laſt year, when Kaſſem was reſtored under the protection of Ruſſia. Muſtapha did not ſubmit to the uſurper Aga Mahommed, who, in December, 1795, on his return from the campaign in Georgia, croſſed the Kurr and penetrated as far as Schamachi. Nor did he truſt in a

defence

defence of the capital againſt Aga, but fled with his party to the mountains, where he remained till February, 1796, when the uſurper retired. The houſes of thoſe who would not follow Muſtapha, (moſt of them were partizans of the Shah's ſect, and attached to Aga Mahommed from religious motives,) were deſtroyed before his flight; and as Aga Mahommed did the ſame to thoſe of the fugitives, the whole city became a ruinous heap. The villages neareſt the plain were equally devaſtated, but many eſcaped by being at a diſtance in the mountains.

Beyond the mountains, where they border on the barren region, there are but few villages; for the natives, particularly between Old and New Schamachi, make a kind of moveable huts of twigs with a ſemi-cylindrical roof, which they cover with felt and ruſh-mats; and thus lead a nomadic life near the rivers and ſtreams. Between the mountains, in the neighbourhood of Old and New Schamachi, there are various villages entirely inhabited by Armenians. In a romantic ſpot near one of them is a convent, with a church built of maſſive hewn ſtone. But the biſhop, with his few monks, is at preſent as poor as the Armenian peaſantry there.

KHANSHIP OF SCHEKI.

The Khan poſſeſſes a diſtrict in the upper mountains beyond the Kurr, on the frontiers of Ganſcha and Georgia. Further on, it is bounded by Leſgiſtan, and on all other ſides by the territory of Schamachi. Nughi is the capital, and ſtands on a ſmall river that falls into the Kurr thirty verſts below it.

View of the History and Geography of the Provinces between the Terek and Kurr, during the first and middle Ages.

In the Grecian and Roman authors, as well as among oriental hiſtorians and geographers, we find but few and general accounts of theſe provinces. In thoſe times, they were ſeparated from the civilized ſtates of Europe by the extenſive poſſeſſions of the rude nations, whom Europeans included under the general denomination of Scythians and Sarmatians. The barren, trackleſs, and hilly countries beyond the Kurr, divided them from the enlightened eaſt; and impaſſable

paſſable mountains, partly buried in eternal ſnows, ſhut them in on other ſides. Thus they were rather barriers to the cultivated nations, particularly the eaſtern, againſt the inroads of Barbarians, than objects of conqueſt; which could have been attained only at the expence of force, and with great danger. In the hiſtory of the Aſſyrio-Babylonian and Median monarchies, we find no traces of thoſe provinces having been ever included. I am therefore aſtoniſhed, that Herbelot and Gmelin could ſay, that they formed the N. W. part of ancient Media. In remote ages they were more probably underſtood by the general name of Scythia. Nor were they ſubject to the Perſian monarchy. Beſides, it is more than conjectural that the unſucceſsful campaign of Cyrus againſt Queen Tomyris, which ended in his death, was undertaken againſt the Caucaſian nations, and certainly immediately againſt the country we are ſpeaking of. After Cyrus had fortunately paſſed the Araxes, (mod. Aras,) and penetrated to its conflux with the Kurr; (Cyrus;) he incautiouſly ſuffered himſelf to be entangled in the defiles, (probably in thoſe where the fortreſs of Ganſcha now ſtands,) and was ſlain with his whole army.

From

From this memorable circumſtance, apparently, the river afterwards was called Cyrus.

Strabo (lib. ix.) explains the origin of it; for he ſays, " the Cyrus and Araxes flow to the E. of the Caſpian; the former through Albania and Iberia, the latter through Armenia:" Again, " Between Albania and Colchis there is a plain watered by various rivers, of which the Cyrus is the moſt conſiderable. Formerly it was called Corus; and it is aſſerted, that Cyrus changed the name and gave it his own."

Even Alexander the Great does not appear to have extended his conqueſts as far as the N. W. part of Caucaſus on the Caſpian. Plutarch ſays expreſsly, in the life of Pompey, that the Iberians, (a nation at preſent inhabiting the greateſt part of Georgia and Kartel,) were never ſubject either to the Medes, Perſians, or Macedonians; becauſe Alexander was obliged to abandon Hyrcania, (modern Gilan and Maſanderan,) too ſoon. Hence it is clear, that he had not time to penetrate to the Caſpian through Hyrcania; conſequently could not have been in modern Schirvan.

Q. Curtius, in his deſcription of the country, is ſo inconcluſive and fabulous, that one

can

can ſcarcely deduce any thing from it. Perhaps his Leuco-Syrians are the ſame with the Albanians. We ſhall ſoon ſhow that the latter inhabited thoſe parts.

We certainly find it aſſerted, by oriental authors, that Alexander firſt fortified the narrow paſs on the Caſpian, at preſent occupied by the city of Derbend; but they are not agreed either in regard to the perſon of this Alexander, (Eſkander in oriental pronunciation,) or to the object; namely, the wall. The majority do not attribute it to the Macedonian monarch, but to a much more ancient hero of the ſame name, who acted the part of a great conqueror in the fabulous hiſtories of the eaſt. Many have even extended the wall ſo far to the eaſtward of Derbend, that Herbelot, in his Oriental Dictionary, aſks, "whether they did not mean the great Chineſe wall?" What oriental writers mention in this reſpect, is ſo contradictory, and diſguiſed under the moſt abſurd fables, that it does not bear the leaſt evidence of hiſtorical truth.

During the ſovereignty of the Parthian and Armenian monarchs, over the chief provinces of the ancient Perſian empire, the nations around Caucaſus were divided into

 three

three ſtates, *viz.* Colchis, Iberia, and Albania; to the latter thoſe provinces indiſputably belonged. But Albania was not ſolely confined to them, for it extended higher up the Kurr, and included part of modern Georgia, on both ſides of that river, as far as the Alazan; yet the mountaineers of Leſgiſtan may partly have been ſubjects of the Albanian monarchs. This kingdom muſt generally have been of tolerable circumference; as it could ſet on foot, according to Strabo and Plutarch, 60,000 infantry and 12,000 cavalry. Of the Roman generals, Pompey the Great was the firſt, and perhaps the only one, who entered that country with an army. In the Mithridatic war, according to Plutarch, after the defeat of Tigranes, he wiſhed to march from Armenia directly over Caucaſus, in purſuit of Mithridates, who had fled towards the Cimmerian Boſphorus. The road carried him firſt through Albania, where a paſſage was granted; but as he wintered there near the Cyrus, probably in that diſtrict of modern Georgia which then belonged to Albania, the Albanians, with 40,000 men, croſſed the river to attack him; but they were totally defeated, and obliged to ſue for peace. Hereupon, Pompey continued his march

march through Iberia and Colchis, and had arrived at the Phaſis, when a freſh inſurrection obliged him to return. After very difficult marches, moſt probably made through the deſerts between the Kurr and Aras, he met, near the Abas, (very likely one of the ſmall rivers that run through the defiles of Ganſcha into the Kurr,) with 60,000 infantry and 12,000 cavalry, headed by Coſis, the king's ſon.

Pompey muſt neceſſarily have paſſed thoſe defiles to penetrate into the heart of Albania, or modern Dagheſtan and Schirvan. It is to be ſuppoſed, therefore, that the Albanians awaited him at the entrance. Strabo obſerves, that they were ſtrongly fortified. Ganſcha ſtands on a rivulet of the ſame name, that falls into the Kurr after a courſe of forty verſts. It is an important paſs between Erivan and Georgia.

The Albanians were entirely defeated, and their leader killed by Pompey's own hand; the king was, therefore, obliged to ſubmit. Afterwards he wiſhed to deſcend the Kurr to the ſea, about three days' journey; but his plan was fruſtrated by the numerous poiſonous ſerpents with which the country ſwarmed; namely, the wide-ſpreading plain of Moghan, on the right bank of the Kurr. He retired, therefore, to Armenia Minor.

In the times of Imperial Rome, the kings of Albania, moſtly as allies of the Iberian monarchs, took part againſt Armenia in the treaties between the Romans and Parthians, of which we have different accounts in the Annals of Tacitus. During the reign of Tiberius, who had given the kingdom of Armenia to Mithridates, brother to Pharaſmanes, king of Iberia, the Albanians entered into an alliance with the Iberians againſt the Parthian monarch Artabanus, who contended the poſſeſſion of Armenia with Mithridates. To this intent, the neighbouring Sarmatians aſſiſted them, as they croſſed Albania to the ſea-coaſt, and broke into Armenia through the paſſes of Ganſcha. The Iberians and the allies were ſucceſsful in that undertaking, and Tacitus furniſhes us with many intereſting particulars of their cuſtoms and method of warfare.

To this Parthico-Armenian period belong the accounts of oriental authors, who ſtate that Jezdegerd, ſon of Baharani, and Nuſhirvan Choſroes, who are reckoned in the fourth dynaſty, reſtored the wall near Der-

bend,

bend, built by Eſkander. Of the latter it is mentioned, that he continued the wall a league into the ſea, and granted a privilege to the governor, in conſideration of his important poſt, of ſitting on a golden throne when he adminiſtered public juſtice. Hence Derbend, in after times, aſſumed the title of *Serir el Dheneb.*

From thoſe accounts it appears, that Schirvan and the adjacent diſtrict, muſt have been independent of Perſia, at leaſt at that period.

I come now to the hiſtory of the provinces, in the middle ages, and ſhall follow the hiſtorians of the eaſt. To be more clearly underſtood, I muſt previouſly mention the denominations attached either to the whole tract of country under diſcuſſion, or to particular parts.

Under the word Schirvan, appears to be comprehended about the ſame diſtrict as at preſent; including the cities of Baku and Shamachi, and a ſtrong caſtle called Calâat el Nagia, or Negia, on the frontiers near Adirbiſhan. Derbend alſo was often reckoned in this province. The part of Dagheſtan lying N. W. of Derbend, and the province of Kumuk, with the greateſt part of N. weſtern Caucaſus, appear to have belonged to the Khozars, whofe poffeffions extended a confiderable way in the plain near the fea of Afov, and nearly as far as the Euxine. We often find this country and Georgia comprifed under the denomination of Gurgiftan; and Schirvan is called by orientalifts, Upper Armenia, (Herbel. Bib. Or.) Some paffages in this author are remarkable, which mention a country there called *Allan*, governed by its own kings. Now when we compare this word with the old name of Albania, we are induced to derive one from the other; although orientalifts do not give the fame extent to the former as to the latter. If we would adopt Herbelot's opinion, which is not improbable, the aboriginal Alans, who were fo confiderable at the time of the migration of the nations to the weft, were only defcendants of the ancient Albanians. The Avars, likewife, originated probably from the fame country; and perhaps the Lefghae tribe of that name, whom we fpoke of before as Dagheftanic borderers, are a remnant in their primary fettlements. From all this it appears, what a diftinguifhed part they took, at the beginning of the middle ages, in the deftinies of Europe. The fovereignty of the Grecian emperors on the Euxine and Afov feas, appears, foon after the migration, to have promoted the introduction of the Greek religion into the weftern parts of Caucafus. The countries on the Cafpian, N. W. of Derbend, were probably in the fame fituation. On the contrary, Schirvan apparently adopted Mahommedanifm very early, and had its own regents or fhahs; confequently cannot have been very independent of Perfia. Under Bathek the ninth Kalif, of the family of the Abaffides, about the middle of the ninth century, Filan Shah of Schirvan is faid to have acknowledged the fupremacy of the Kalifate; and Manugeher Shah, who refided in Schamachi in the middle of the twelfth century, was efteemed a patron of the fciences. The middle ages, namely, from the introduction of Mahommedanifm down to Dfchengis Khan, appear to have been the grand period of opulence in Schirvan.

It was quite otherwife in the country beyond Derbend, then inhabited by the Khozars. Under the Kalifs they often made inroads into the fouth-eaftern provinces of Afia. Abdalmalek, the fifth Kalif of the family of the Ommiades, who reigned about the end of the feventh century, undertook a campaign againſt them, defeated them near Derbend, and obliged thoſe who eſcaped the ſword to be circumciſed. But their ſubjection does not ſeem to have been of long duration; for, at the commencement of the eighth century, under Kalif Heſcham, his general, Mervan, reconquered the country, and penetrated far into the territory of the Khozars. Towards the cloſe of the the ſame century, we ſee the Khozars again, under Harum el Raſchid, iſſue from the mountains and carry off nearly 100,000 priſoners, (Herbel. Bib. Or.) They afterwards eſtabliſhed themſelves in the country; and hiſtorians ſay, that they were firſt ſubdued by the Shahs of the Seldſchuk dynaſty, who flouriſhed from the beginning of the tenth to the end of the eleventh century.

After the great revolution effected by Dſchengis Khan and his ſucceſſors, the provinces acknowledged the authority of his family, to whom Turkeſtan and Perſia had devolved. During the ſubſequent commotions amongſt his deſcendants, they ſuffered greatly from the incurſions of the princes who reigned in the north-weſtern parts of the Caſpian. In the beginning of the government of Abaka, the eighth Mongul

prince

prince of Dſchengis's family, who reigned to the ſouth-eaſt of the Caſpian, Barcah, a deſcendant of Giagathai, of the ſame ſtock, and who reigned in the north-weſt, wiſhed to penetrate into Perſia. But he was defeated near Derbend in 1265, and obliged to retreat. However, he made another attempt at the head of 300,000 men, took Derbend, and penetrated as far as the Kurr. But he died juſt at that moment, and then his army diſperſed and retired. In 1320, the Shah Uſbeck, from Kiptſchak, broke in on the ſame ſide, and made himſelf maſter of Derbend. Abu-Said, then Sultan of Perſia, marched an army to the Kurr with a view of defending the paſſage of that river. As Uſbeck was engaged in plundering the flat country, Abu-Said had time to gain conſiderable reinforcements, ſo that the Shah was obliged to retire; but he could not ſecure his retreat without great loſs. In 1335 he renewed the campaign, and Abu-Said, the following year, marched to Schirvan to oppoſe him, but was carried off by ſickneſs. After his death, Uſbeck appears equally to have been impeded in his expedition. (Herb. Bib. Or.)

Towards the cloſe of the fourteenth century, when Timur-Lengh, (known in Europe

by

by the name of Tamerlane,) extended his conqueſts as far as the weſtern coaſt of the Caſpian, Emir Ibrahim, then regent of Schirvan, willingly ſubmitted to him. Hereupon Timur marched through that province and Dagheſtan againſt Toktamiſch, lord of Kiptſchak, and obliged him to ſeek refuge in the mountains. But from thence, by the aſſiſtance of the Khozars, he made many irruptions into Georgia and Armenia, until Timur, a ſhort time before his death, put an end to thoſe expeditions. In 1410, Emir Ibrahim broke into Adirbiſchan, and took the city of Tauris, and in the following year was maſter of Iſpahan; but he loſt his conqueſts before the end of it.

At the concluſion of the fifteenth century, Ferokzhad was Shah of Schirvan, who was vanquiſhed and ſlain by Iſmail Sophi, in 1500; and by this conqueſt he began the ſubjugation of all Perſia. His family poſſeſſed thoſe provinces, nearly uninterrupted, until Peter the Great made himſelf maſter of part of them. Schirvan and Dagheſtan, during the government of the Sophi, were more than ever dependent on the crown of Perſia; and the wiſe adminiſtration of Shah Abbas the Great had very beneficial effects. The caravanſerais,

vanſerais, bridges, and wells, whoſe ruins we ſtill meet with on the high roads, are moſtly to be attributed to his happy government.

GEOGRAPHY OF THE FIRST AND MIDDLE AGES.

We obſerved that the provinces above mentioned conſtituted the principal part of the kingdom of Albania. The branch of Caucaſian mountains belonging to it, at leaſt the north-weſtern part, was diſtinguiſhed by the name of the Ceraunian.

Beſides the rivers Cyrus and Alazonius, there are four particularly named by the Greek and Roman geographers, that fall into the Caſpian. They are enumerated in order from N. W. to S. E. namely, Soana, Gerrhus, Caſius, and Albanus. Strabo, Pliny, and P. Mela, mention likewiſe the Cambyſes, whoſe preſent poſition I cannot determine. The paſſage in Mela, (lib. iii. cap. 5.) appears to confound it with the Araxes. The firſt flowed near the frontiers of Sarmatia, and is therefore, probably, the modern Terek. The Gerrhus can be no other than

than the Koiſu. The Caſius is perhaps the Manaſſa, the moſt conſiderable of the Dagheſtanic rivers; and the Albanus is the preſent Samur.

There is certainly ſome ſimilitude between the names Caſius and Koiſu, ſo that both might be taken as one; but then the frontiers of Albania muſt neceſſarily be extended beyond the Terek, very near to the Kuma; and that in other reſpects could not be the caſe.

The following cities are ſtated by Ptolemy: Between the Soana and the Gerrhus, Teleba on the ſea; inland, Thalbis. Between the Gerrhus and the Caſius, on the Gelda, Thabilæa in the country of Thiauna. Between the Caſius and Albanus, Albana on the ſea; inland, Chadaca, Miſia, Bozicata, Chabula, and Chobata. Between the Albanus and Cyrus, Gagna on the ſea; and in the neighbourhood Baruca, Sioda, Oſica, Abdiabda, Abliala, and Comechia. Higher up in the mountains, Jobula, Juna, Moſega, and Samunis; further on, between the Cyrus and another river that falls into it, not named, perhaps the Akſu, Niga, Dechlana, Bacchia, and Tetragda. Of all theſe, through the lapſe of ages, and numerous revolutions, to which

which the country has been expofed, not the flighteft traces remain; it is difficult, therefore, to hazard a conjecture with regard to their fituation. Teleba probably ftood between the mouths of the Terek; Albana on the fite of Derbend; Gagara, where Baku now is; and Comechia on the fite of Schamachi.

We meet with very contradictory accounts of that famous pafs, which feparated the northern nations from the eaftern. In comparing the various paffages in the Greek and Roman authors, it appears there were properly three. The *Portæ Ibericæ* croffed the mountains beyond the frontiers of Albania, perhaps nearer the fource of the Terek, behind Mofdok; the *Portæ Albanicæ* probably lay in the mountains not far from Derbend; and the *Via Cafpia* was properly that narrow pafs between the mountains and the fea now occupied by Derbend.

The country is principally divided into provinces by oriental geographers of the middle ages; namely, Schirvan, Serir el Dheneb, and the part belonging to the Khozars' territory.

Schirvan, by far the moft important and beft cultivated, had nearly the fame boundaries

daries as at present. Its capital, (Herbel. Bib. Or.) lay in 40°. 50′. N. lat. a proof that Old Schamachi ſtands exactly on its ſite. It was the reſidence of the regents, and appears, from the times of the Kalifate down to Dſchengis Khan, to have attained to no common degree of civilization and opulence, according to eaſtern manners. It gave birth to Felekir, one of the moſt celebrated Perſian poets, who lived under the regency of Manugeher Shah, and died in 577 of the hegira, (Herbel. B. Or.) It is ſingular, that different writers place this city on the Caſpian, and make it a ſea-port. Perhaps they fell into the error by the circumſtance that, as Schamachi was ſo famed for being the ſtaple between Perſia and the north, the port of Baku was thereby overlooked.

Being placed in 39°. 30′. N. lat. by oriental geographers, it agrees very well with the ſituation of modern Baku. It was celebrated for its harbour in the middle ages, though leſs known than Schamachi. Beſides the name of *Deria* Baku, (the ſea of Baku,) the Caſpian had others.

Nagia, or Calâat el Nagia, was a caſtle celebrated on account of its uncommon ſtrength. As it lay near the frontiers of

Adir-

Adirbifchan, it was probably on the Kurr. In the middle ages it ftood many long fieges, and was generally confidered as one of the ftrongeft forts in Afia. At the confluence of the Kurr with the Aras, we fee the ruins of a fortrefs deftroyed by the prefent Perfian ufurper Aga Mahommed. It is called Kalagail, and belongs to a Bey, defcended from the Khans of Ardevill, who is proprietor of the village of Dfchawat, directly oppofite, on the left bank of the Kurr. It is not improbable, that it is the ancient Calâat el Nagia.

The province of Serir el Dheneb included only the city and territory of Derbend. How far the latter extended on both fides of the city, as well as towards Schirvan and Daghestan, is not mentioned by hiftorians. Of Derbend we have fpoken above. They mention a city and road-ftead called Tekki, or Tarku, not very far from Baku. Perhaps by Tarki may be underftood the modern capital of the Schamgal. But, though the city does not ftand at prefent on the Cafpian, yet veffels can anchor in front of it. Probably this Tekki was the old Terki at the mouth of the Terek, whofe inhabitants Peter the Firft removed to Kiflar.

Physical

Physical Properties and Productions.

CLIMATE.

In a country, lying nearly between the 39th and 44th degrees of N. lat. we naturally expect a very mild climate; rather oppressed by heat than severe cold. It is, however, less hot than in parallel latitudes of southern Europe, which is to be attributed to its situation near the sea, and to the north of one of the Caucasian chains, whose summits are buried in eternal snows. A consequence of this position is its peculiar inconstancy, which affects the health of strangers unacquainted with necessary precautions. In this respect it is very similar to the northern part of the Taurid.

At the end of March, (I follow the old Russian style,) and the beginning of April, spring sets in; and even at the close of the latter month the heat begins to be oppressive, which continues through May, June, and July, and is at the highest in the dog-days. According to my own sensations, (for I had no opportunity of making thermometrical observations,) it is not greater than in the

southern

ſouthern provinces of Germany, that lie ſome degrees more north, except in the neighbourhood of Baku and the Kurr. It is daily mitigated by ſea or land breezes; and the nights are cool, even in the hotteſt ſeaſon, although the mountains are above 100 verſts diſtant. There are partial ſhowers in ſummer, at leaſt on the mountains; but ſtorms ſeldom happen; and the atmoſphere diſcharges itſelf without thunder and lightening, even at the higheſt degree of heat. The cauſe I cannot preciſely aſcertain. Perhaps exploſions take place in the upper mountains, ſo that the equilibrium of the electric fluid is in moſt caſes reſtored before the clouds deſcend.

Towards the autumnal equinox violent ſtorms, with changeable weather, ſucceed, though there are always intermediate days of heat. At the cloſe of September, and in October, the rainy ſeaſon begins, and then the middle mountains are white with ſnow. Even in November the nights are often froſty, and moſt trees do not ſhed their leaves before the end of the month. In December we have froſts and ſnow-ſtorms, but the ſnow very ſeldom lies more than two days in the plain. In many ſeaſons there is ſpring-weather in January, ſo that the trees begin to bud and ſpring-flowers to put forth. But, as the country is not protected from the north winds, it is only a tranſient effort of nature reviviſcent; there being frequently more ſevere froſts and ſnow-ſtorms from February to the end of March, than in November and December. The weather is rather changeable till April, when ſpring commences. It is to be underſtood, however, that different ſituations have different modifications. The higher we aſcend the mountains, the more temperate is the air, and dew and rain are more common in ſummer, and froſt and ſnow in winter; on the contrary, the lower mountains and plains are proportionably expoſed to a greater degree of heat and drought. Except on the mountains, the air is rather prejudicial to health; but by precaution, and means of guarding againſt ſudden changes and autumnal fogs, the danger is conſiderably alleviated. We have the example of the natives, who do not lay aſide their peliſſes even in the hotteſt ſeaſon. Certainly there are ſituations where circumſpection would be of little avail, particularly in the marſhes near the Kurr, where mephitic vapours muſt ne-

ceſſarily

ceſſarily ariſe in hot ſummers. In the neighbourhood of Baku ſtrangers are ſubject to cutaneous diſorders, even after a very ſhort reſidence. This may very well proceed from the atmoſphere being impregnated with ſulphureous and bituminous particles, but not conſequently injurious to the conſtitution.

PRODUCTIONS.

My obſervations, on this head, relate principally to the province of Schirvan, as I only curſorily paſſed through Dagheſtan and Kumuk. By what I can judge from my own information, and from that of others, the countries are not eſſentially different; ſo that my remarks, in the main point, may apply to all.

GEOLOGY.

The chain of mountain between the Terek and Kurr, is a branch of thoſe Alpine ranges in the neighbourhood of the Black Sea, which are to be conſidered as the heart of the great Caucaſian groups. From the higheſt point of theſe Alps, (the Ellborus,) to the north-weſtern commencement of the above chain, is above 500 verſts. If we wiſhed to give a ſyſtematic orography of the latter, their ſummits ſhould neceſſarily be inveſtigated. But that is impoſſible, as they are inhabited by wild Leſghaes, who will not ſuffer any European to approach them. But ſhould time and circumſtances become more favourable for ſuch examinations, then, the obſervations which I am now going to make, may claim their place in a complete deſcription of Caucaſus.

All that I have ſeen of the mountains belong neither to primitive, nor to ſecondary ones; but are alluvions, which I ſhould not rank very high among the revolutions of our globe: although they have apparently undergone great changes ſince their formation. It is probable, that the ſame holds good with regard to the ſummits; for, I have approached them in different places within four or five leagues, without perceiving any indications of a neighbouring primitive chain under the *ſtrata*. The inhabitants, particularly near Schamachi, fetch their millſtones, which are of the granite genus, from the Araratian mountains between the Kurr and the Arras; a proof that thoſe mountains contain no ſuitable kind of ſtone. The diſtance

diſtance of the Caucaſian Alps, and the circumſtance of this chain lying neareſt to the Caſpian and the wide-ſpreading plains to the N. W. whoſe ſoil is argillaceous and ſaline, in ſome degree aſſiſt our ideas reſpecting their recent formation. I have no where ſeen traces of granite, gneiſs, horn-ſchiſt, or volcanic productions.

The neareſt promontory to the ſea conſiſts of thick *ſtrata*, almoſt horizontal, of lime-ſtone full of ſmall muſcles, moſtly of the chama kind. We ſcarcely meet with any extraneous ſubſtance amongſt them; the ſtone, therefore, appears porous, but is ſufficiently hard for building, as is evident from the walls of Derbend, which are entirely built of it. Between the Rubas and the Atatſchai, that range of calcareous mountains is interſected, and does not appear again till near the peninſula of Abſcharon, from thence it extends uninterrupted to the coaſt, and dips towards the Kurr.

The lime-ſtone often ſhows itſelf, on the declivities, in ſcattered maſſes of uncommon forms, particularly on the road between Baku and Sallian.

Behind the calcareous range, there is one of ſlate, which is by far the moſt extenſive.

The *ſtrata* generally decline gradually from W. to E. and have uniformly a ſtrong inclination to the plane of the horizon, with which they often form an angle of more than fifty degrees; the point of inclination being to the northward.

This range is full of precipices and narrow dells, worked out by the torrents acting upon the ſoft argillaceous and marl *ſtrata*. The latter are moſtly foliated, whereby they differ in degrees of hardneſs, weight, fineneſs, and colour, *ad infinitum*. They often contain organic remains, particularly in the neighbourhood of lime-ſtone. The colour depends principally on the admixture of iron; hence the different ſhades of red and yellow prevail, which ſometimes change into brown and black. Aſphaltum often forms the baſis of the latter, eſpecially in the diſtrict of the bald mountains, which we ſhall ſpeak of hereafter. Many of the ferruginous argills are real ochres, but whether they contain ſufficient iron for ſmelting appears very doubtful.

Here and there we ſee, likewiſe, layers of a bright green colour, containing thin veins of milk-white ſpath, which run in a parallel direction. The latter are commonly leſs

foliated

foliated and harder than the upper ones, and ſhow themſelves by their great ſpecific gravity. I have often found them impregnated with an extraneous ſubſtance, that tinges them with violet and black. According to all eſtabliſhed proofs, they appeared to conſiſt of argill and calcareous earth with carbonic acid, beſides iron and aſphaltum. Porcelain and pipe clays, and fullers'-earth, are partially to be met with. Between the ſoft and foliated *ſtrata* are layers of a harder ſpecies of ſtone, that contain nearly ſimilar conſtituent parts. The city of old Schamachi is entirely built of it. The mountain of Peſch-parmak is compoſed of the ſame ſtone, though more calcareous; and has a yellow appearance, from being mixed with a ſpecies of tripoli. Flints and ſelenites are alſo partially diſtributed. Iron and pyrites are not in that abundance we ſhould have expected in mountains of ſuch a nature. The higheſt ranges which I could examine, particularly thoſe in front of the plain between the Rubas and Atatſchai, are compoſed of ſandſtone, fine-grained, effervefces a little with acids, and appears to contain moſtly argill and ſilex; on that account, the rivers which flow from them are very muddy. I have alſo remarked a ſpecies of red marble with white veins, but it does not take a high poliſh.

The elevated region, beyond the real chain, merits particular attention. The mountains there belong moſtly to the ſchiſtoſe claſs, and are ſhut in by calcareous ones towards the ſea. They are peculiarly diſtinguiſhed by a rich combination of ſulphate of lime, muriate and ſulphate of ſoda, and ſulphate of magneſia, but eſpecially of aſphaltum. The ſulphate of lime lies between the *ſtrata* in numerous ſcattered fragments, like talc; likewiſe, in ſmall ſemi-tranſparent cryſtals; and frequently in long ſlender aggregated cryſtals, forming a coating of ſcarcely the thickneſs of a ſheet of paper.

The ſchiſt itſelf is ſometimes very ferruginous, at others bituminous, and then more or leſs of a black colour. We do not find it merely near the ſea, but to a great extent as far as the higher ranges, partly impregnated with ſemi-fluid petroleum, and partly ſhowing its union with aſphaltum by the ſmell. It is no wonder, therefore, that ſo many ſprings of petroleum and naphtha ſhould riſe on the ſea-coaſt of this diſtrict, particularly in the peninſula of Abſcharon,

that

that ſtretches fartheſt into the ſea. The ſalts, in combination with water, appear moſtly adapted to exclude the aſphaltum from thoſe mountains which contain it; and the naphtha and petroleum, flowing from the ſchiſtoſe mountains, collect in the lime-ſtone of the promontory. There too we meet with the greateſt number of ſalt lakes. The ſulphates of ſoda and magneſia are partly combined together, and partly united with muriate of ſoda. The firſt univerſally predominates. In many places it is found coating argill in the form of white tufts or powder, or depoſited on the edge of the ſprings, where it was held in ſolution. In ſhort, it is nearly of the ſame nature and properties as the Aſtrachanic ſalt uſed in Ruſſia.

On ſome ſpots we meet with ſmall conical hills, that throw out air-bubbles and muddy water. They are perfectly of the ſame nature with thoſe in the Taurid, ſo circumſtantially deſcribed by Profeſſor Pallas.

I think it proper to make ſome obſervations on the remarkable inflammable part of the peninſula, in reference to Gmelin's deſcription.

His repreſentation of the place is correct; the leſs ſo, on the contrary, is his manner of of explaining that ſingular phænomenon. According to his account, it is an earth impregnated with naphtha, having the peculiar property of taking fire. But I had an excellent opportunity of being convinced, that the inflammable matter is an aëriform ſubſtance. The inhabitants brought leathern bottles full of it to the Ruſſian camp at ſome leagues diſtant, with which very ſimple proofs, but perfectly concluſive, were made, in the preſence of a number of witneſſes.

When we ſtuck a reed into the mouth of one of the bottles, the aëriform ſubſtance immediately caught fire, and burnt with a ſteady, or ſtrong flame, according as the confined air was more or leſs preſſed towards the mouth; but no ſmoke or unpleaſant ſmell of naphtha were perceived. Now ſhould be enquired, whether it differs from common inflammable air, or not. In the firſt place, modern chemiſts will ſcarcely previouſly conjecture, or be inclined to agree with me, that this inflammable gas is nothing but hydrogen. To confirm the opinion a circumſtance muſt be added, which is, that the inflammable air confined in the earth catches fire only after the ſoil has been turned up ſome inches in depth; for, in this

caſe,

case, a column of air, mostly composed of that substance, is formed; and in every other the confined gas evaporates too rapidly, on account of its specific lightness. Even Gmelin might have been reasonably convinced of the nature of this substance, from what he had observed in the Indian temples there, where reeds can be lighted or extinguished at pleasure. Perhaps he was prevented by the circumstance that, in his time, the theory of the various gases, which we owe to Lavoisier and his successors, was not so clearly developed.

To close this section, I must say a few words with regard to the plain near the Kurr.

Its locality, as well as the properties of the soil, are convincing proofs of its recent formation by the retrogression of the sea. A passage in the Annals of Tacitus, (lib. vi. cap. 33.) appears remarkable in this respect, for it states, "that the pass between the sea and the exterior Albanian mountains, (which was where the mountains of Schirvan stretch into the plain near the Kurr,) was impassable in summer, on account of the encroachment of the sea." As that is not the case at present, it follows, from the above

passage,

paſſage, that the ſea muſt have retired conſiderably ſince the time of Tacitus.

The *ſtrata* are completely vertical, and we frequently find large beds of muſcles, ſuch as are moſt common at preſent in the Caſpian. They are very little changed, having only loſt their enamel; they are not even petrified, but lie imbedded in argillaceous marl with looſe pebbles. There are alſo about the plain, and in the channels of the torrents, rounded fragments of various kinds of ſtone, which are no where elſe to be met with; conſequently muſt have been brought by the waters.

In ſhort, the ſoil is generally argillaceous, and mixed with ſulphate of lime and muriate of ſoda.

PLANTS.

Vegetation is uncommonly diverſified, and offers a wide field to the botaniſt for intereſting reſearches and diſcoveries. Whoever is acquainted with the plants of the Taurid will find, that theſe are of a very ſimilar nature, although there are many ſpecies which would be ſought for in vain in the Taurid. Amongſt them I have obſerved thirty-three,

not

not one of the Cryptogamia claſs, which have been hitherto non-deſcripts, or at leaſt too curſorily mentioned in old botanical catalogues, to be admitted into the ſyſtems of the moderns.

Their definition, with a ſhort deſcription and obſervations on ſome other ſcarce ſpecies, will be found in the Appendix. One of the new ſpecies, (*mimoſa Stephaniana,*) I have named after Profeſſor Stephan in Moſcow.

The moſt common foreſt trees are oaks; white beeches; elms; wild prunes; apples; and two kinds of pears, namely, the common, and the *pyrus ſalicifolia.* When one conſiders, that fruit-trees thrive beſt grafted on parent ſtocks, *viz.* apricots on prunes, peaches on almonds, and apples on quinces, it would be worth the trial, whether our finer ſorts of pears would not be improved by being grafted on oriental ſtocks. Leſs common are the red beech, common and manna aſh, Linden cherries, cornelian cherries, quinces, medlars, all-ſpice, and ſervice trees. The *elæagnus anguſtifolia* and *orientalis* are tolerably abundant. Its ſweet mellow fruit, which has the form and nearly the colour of the date, is ſold in the markets of

Baku,

Baku, Schamachi, and Derbend. The Tſchinar *(platanus)* much prized here and in Ruſſia for its wood, ſeldom appears but in thick foreſts. The juniper, of the ſavin genus, (indigenous alſo to the Taurid,) grows commonly among the rocky acclivities expoſed to the ſun, of conſiderable height and thickneſs. I conſider this ſpecies as differing from the ſavin, and ſhall ſpeak of it in the Appendix, as the *juniperus excelſa*. The common juniper abounds in dry, ſtony places, and on the hills; but the yew-tree is ſcarce. Birches and firs we never meet with. Cheſnuts grow near the ſummits; the fruit is ſold very cheap every where in the month of October. The ſilver and black poplar, the alder, and various aquatic ſorts common in Europe, grow in moiſt places. The firſt conſtitute the great part of the wood along the banks of the Kurr. The Italian poplar here, as in many other places in the eaſt, ſerves to ornament gardens, and public ſquares in cities and villages. I have generally met with the terebinth and piſtachio on the heights along the Ata, but not in thick foreſts among the mountains.

The largeſt plantations of granates are on the ſides of the mountains oppoſite New

Schamachi,

Schamachi, and on the borders between the woody and bald mountains. In their natural ſituation they ſeldom grow to the height of a tree. The fruit ripens in October. There are ſweet and ſour ſorts; the laſt are the commoneſt, and diſtinguiſhed by their ſmallneſs and dark-red colour. Black and white mulberries grow wild on the Kurr; in other places they are ſeldom found wild, being probably only runners from the plantations. In the foreſts and among the buſhes the vine is uncommonly abundant, eſpecially in the plain between the Rubas and the Ata, where it runs up to the higheſt branches of the trees. Figs and walnuts are partially ſcattered among the buſhes, but too ſparingly to aſcertain whether they are really wild.

The moſt common ſhrubs are, whitethorns of three ſpecies, one being the rough oriental; bullace, ſloe, hazle, common hip, ſpindle, black elder, purging buckthorn, and evergreen buckthorn. The latter grows every where on the acclivities; and, on account of its prickles, is not improperly called by the Ruſſians *catchbuſh*. The ſmall-leaved dwarf buckthorn, indigenous to the country of the Grebinſki Koſaks higher up the Terek,

is

is very common, and might be advantageoufly employed for tanning, as the fumach with egg-fhaped leaves is ufed in the preparation of morocco. The *lycium ruthenicum* grows partially on dry fpots expofed to the fun. The *hippophaë rhamnoides* thrives in ftony places near the rivers, and makes a pretty appearance from its fcaly leaves and yellow berries. The *mefpilus pyracantha*, and the round-leaved, are frequently to be met with in bufhes. I have feldom feen the real honey-fuckle. The *jafminum officinale* grows amongft the ruins of Old Schamachi, probably from the gardens formerly there. The *jafminum fructicans* is common in ftony elevations open to the fun. The *colutea arborefcens* grows partially in ftony valleys. The French tamarifk fhows itfelf particularly in the plains; in woods it is more rare, but generally of the height of a tree. It would be worth while to make plantations of it in the barren and faline deferts in the fouth-weftern parts of Ruffia, as this fhrub appears to thrive beft where no others will grow.

The *ephidra monoftachia* grows to an uncommon fize, the ftem being often of the thicknefs of one's fift and four feet high. It is a ſtrong ſtyptic, and ſaid to be uſed with great ſucceſs in the dyſentery. There is a ſpecies of laurel in the woods, on the upper mountains, which is probably the *Daphne Caucaſia* deſcribed by Pallas in his *Flora Ruſſica*, from Gueldenſtædt's papers, without having ſeen it. There are various ſpecies of roſes in buſhes and hedges; and in the delightful neighbourhood of Kuba, I have met with, at the beginning of June, the wild *roſa centifolia* in prime bloom. Brambles are common in barren places, and bear flowers and berries to the end of autumn. In the ſunny and ſtony parts of the mountains we generally find the dwarf tragacanth, which Tournefort firſt carried to France from the Archipelago, and is diſtinguiſhed in Lamark's botannical Encyclopædia by the name of the Cretan *aſtragalus*, from various ſpecies included by Linnæus under the general denomination of *aſtragalus tragacantha*. The cells in the roots are full of gum, that exudes in ſpring, and conſtitutes the well-known gum tragacanth. The ſhrub, deſcribed and drawn by Gmelin as the real China *arbor-vitæ*, is certainly the *ſmilax excelſa*. It grows in the buſhes between the rivers Rubas and Samur; but I have not met with it elsewhere.

Whole districts, particularly near Baku and in the plain by the Kurr, are covered with dwarf marine plants of the *salicornia*, *salsola*, and *anabasis* tribes. The Caspian *salicornia* reaches from six to eight feet in height. From these plants carbonate of soda is extracted. The following, likewise, which grow in the neighbourhood, might be used: *salicornia herbacea, Caspica; polycnemum triandrum, oppositifolium; chenopodium maritimum; salsola fructicosa, glauca, vermiculata, hyssopifolia, prostrata; cali rosacea, altissima, muricata; anabasis aphylla; echinus reaumurea, vermiculata; aster tripolium.*

There is no want of wild roots, which are, or might be, used in domestic œconomy. Many are cultivated in gardens, in other parts of the world, as vegetables. For example, celery is common in moist, open places. Fennel is found in the greatest abundance in the vicinity of the more elevated ranges, where it grows to above the height of eight feet. An idea may be conceived of the quantity, when it is mentioned that our army, encamped below New Scha-

machi,

machi, in the autumn of 1796, ſcarcely uſed any other fire-wood for fourteen days than the dry ſtalks. Parſnips we meet with here and there; as well as loveage *(smyrnium perfoliatum)* ſo much liked in Italy and in the ſouth of France. Aniſe is in great plenty on the ſtony banks of rivulets. Various ſpecies of onions and garlicks might be employed for culinary purpoſes. In ſpring, abundance of aſparagus may be cut. It certainly is not the common garden ſort, yet is not inferior in quality, and is difficult to be diſtinguiſhed when firſt it appears. In the Taurid alſo, and even in the ſouthernmoſt parts of the Ukraine, this ſpecies is indigenous, but nevertheleſs has never been properly deſcribed or repreſented by botaniſts. In the Linnæan ſyſtem it is merely indicated as the *aſparagus verticillaris* of Tournefort. The caper tree is ſo common in heavy ſoils, and in ſunny places, that abundance of buds might be gathered. Hops alſo grow in the hedges along the ſtreams and rivulets.

In an œconomical point of view, trials could be made of the different kinds of hemp that grow wild in theſe countries. On hills expoſed to the ſouth, and before in the Taurid, I have obſerved a ſpecies of wheat very ſimilar to the cultivated kinds, and even in its natural ſtate produces very fine corn. In a work, which I am inclined to publiſh, on the plants of the Taurid, I ſhall deſcribe this graſs under the name of *triticum pubeſcens.* It would be worthy of inveſtigation to aſcertain, how near this non-deſcript would approach to the well-known domeſticated ſpecies by culture. Madder is to be found almoſt every where, but in no part ſo fine, and in ſuch plenty, as in the territory of the Uzmey in Dagheſtan. The inhabitants dig up the roots, dry them, and ſell them to Armenian merchants, who export them in great quantities, partly to Kiſlar, and partly by water to Aſtrachan. The moſt opulent of thoſe merchants in Derbend are ſolely indebted to the trade in this produce for their wealth. The poud in Kiſlar coſts from twelve to fifteen roubles. Beſides the red, there are various other roots for dyeing; namely, woad, *reſeda luteola,* dyers' camomile, different ſpecies of the bugloſs genus whoſe roots dye red, *(onoſma echioides, echium rubrum,)* and the *croton tinctorium,* &c.

The following wild plants might be introduced in flower-gardens: the *crocus ſpeciosus,* ditto *autumnalis, vernus,* and *luteus,)* which I ſhall deſcribe in my *Flora Taurica.)* Some kinds of iris, ſcabiuſes, yellow narciſſuſes, tulips, *lyſimachia orientalis* of La Mark, *apocynum venetum,* different ſpecies of wild cloves, larkſpurs, Hyrcanian germander, *phlomis laciniata,* balm-gentle, *heſperiſtriſtis,* ſingle matronalis, yellow autumnal roſe, aſters, *centaurea Babylonica, ſplendens, reflexa* of La Mark, and others. A ſpecies of the wall-flower that approaches to the common ſort, but is particularly diſtinguiſhed by its leaves being deeply indentated, and by the perfume of its brown-red flowers in the night, merits a peculiar ſituation in European pleaſure gardens, and would probably ſucceed as well, by proper attention, as the common wall-flower. This ſpecies is alſo found in the Taurid, and called by Pallas the *cheiranthus odoratiſſimus.*

On the mountains and in the plain, in the Kubaic territory, there is abundance of paſturage. Beſides many of the beſt graſſes, I have met with eight different ſpecies of clover, four of ſnail trefoil, and many kinds of tares and *hedyſarum onobrychis:* various ſpecies of thyme and betony grow on the mountains, with many other common aromatic plants, which afford the beſt food for ſheep.

Medicinal plants are likewiſe plentiful; namely, different ſpecies of veronica, ſages, plantain, pſyllium, valerian, red lythrum, thorn apples, gentian, alkekengi, orpine, centaury, Vienna hemlock, wall-wort, dittany, germander, mint, many *antiſcorbutica* of the genus with cruciformed flowers, marſh-mallows, two ſpecies of liquorice in great abundance, St. John's wort, ſuccory, dandelion, the true elicampane, coltsfoot, camomile, ariſtolochy, ſago-wort, real and ſtone-burnet, various kinds of the common milfoil, *momordica elaterium,* and many ſorts of worm-wood.

The inhabitants attribute to the dwarf worm-wood, *(artemiſia humilis,)* which grows moſt commonly in theſe countries, and in the deſerts between Kiſlar and Aſtrachan, dangerous, and often deadly effects, that it has on horſes. But I am perſuaded it is a prejudice, for in other places, where this plant is not leſs general, thoſe effects are unknown. That the prejudice has exiſted many years, is proved from Bell of Antermony's travels, where, in the deſcription of the campaign of Peter the Firſt, it is mentioned,

tioned, that 500 horſes died in one night from having eaten worm-wood.

ANIMALS.

Of the quadrupeds many are peculiar to warm latitudes. Of the carnivorous claſs we meet with the following: the panther, a particular ſpecies of ſmall lynx, the wild cat, bears, the hyæna, wolf, jackall, and fox.

The panther, (of Buffon,) common in the interior of Perſia, only appears on the banks of the Kurr. Yet, probably, it is not indigenous, but ſtrays from the Gilanic and Taliſchumic foreſts into the wide arid plains, through which the Kurr flows, and takes refuge in thoſe parts of the banks covered with reeds and underwood, where it never wants for food. The dead panthers I have ſeen were, at moſt, ſeven feet long, from the ſnout to the tip of the tail, *i. e.* four in the body and three in the tail. Their form, upon the whole, had nothing remarkable; except that the fore feet were uncommonly thick and diſproportioned to the reſt of the body. The noſe is blunt, the eyes ſmall and fiery, the ears ſhort and not pointed. The whole upper part of the ſkin is of a browniſh light-yellow, darkeſt on the ſpine, becomes lighter on the ſides, and at laſt loſes itſelf in the white of the belly. The noſe is not ſpotted; the whiſkers are white; and the ears have a central, black, oblique ſtreak up to the points, and the edges, on one ſide, are of the ſame colour. The ſpots on the head and feet are ſmall, without any regularity. Along the ſpine they begin to form into rings, which become more viſible on the ſides and towards the belly, and are largeſt and moſt regular on the hind quarters. In fact, they are more ſtriking here, for the centre of the rings always preſerves the deeper colour of the ſpine, whilſt the ground-colour always becomes paler. The neck, breaſt, and under part of the body, are beautifully white, with large black ſpots. The tail above is ſpeckled like the back, but white underneath; towards the point are two or three large ſpots almoſt circular. The males differ little from the female here deſcribed. They are ſmaller, and the ſpots always more irregular, particularly in cubs. But the black mark on the ears and tip of the tail I have found common to all. The Linnæan definition of this ſpecies, *felis pardus; caudâ elongatâ, corpore maculis*

ſuperi-

ſuperioribus orbiculatis, inferioribus virgatis, is not the beſt; and gave Buffon cauſe to doubt, whether his panther was the ſame as the *felis pardus* of Linnæus. In my opinion, the ſpecific characters would be better deſcribed thus: *Felis pardus; caudâ elongatâ apice annulatâ, corpore maculis dorſalibus obſoletè lateralibus diſtinctè orbiculatis majoribus, auriculis brevibus faſciâ nigrâ.*

The panther is an uncommonly ſtrong and active animal, and eaſily ſeizes wild boars, which conſtitute his chief food in this country. He alſo preys on horſes and cattle, but does not readily attack man. The inhabitants call him *paleng*, anſwering to the word tyger as commonly uſed, but they indiſcriminately give this name to other large ſpotted or ſtreaked animals of the ſame genus.

The Caſpian lynx, *(felis chaus,)* firſt deſcribed by Gueldenſtædt, is not unfrequently found in marſhes and poplar-woods on the Kurr, but ſeldom elſewhere. He has perfectly the form of the common lynx, but ſmaller, yet almoſt twice the ſize of the wild cat. The ears are not ſo ſharp-pointed, with black hairs at the tips. The upper part of the body is not ſo red, for the hairs

are

are white at the roots, then black, red higher up, and black again at the points. The insides of the legs are, likewise, rather black, particularly of the front ones, and the pats are perfectly black. The belly is pale reddish, sinking into white, without any mixture of black. The lips, beard, and under part of the snout, are white. The tail is short, and the rest of the body similar to the common lynx.

Bears and wolves inhabit the forests on the upper mountains. Hyænas are sometimes seen in the neighbourhood of the Kurr, and are probably stragglers from the southern provinces, like the panthers.

Jackalls are uncommonly numerous, and prefer the vicinity of inhabited places, where they meet with more dead bodies, on which they feed; on that account, they abound most round Derbend, Baku, Schamachi, and Sallian, where they howl about in packs during the night. Their howl cannot be better compared than to the piercing and continued shrieks of a posse of boys, who meet with some unexpected object to excite their attention. A further account of this animal would be superfluous, as it is particu-

larly

larly deſcribed in Gmelin, though badly repreſented.

The fox is very common. Its colour often approaches to the brown, and hunters ſay that they have met with black ones.

The foreſts are full of deer; and numerous wild boars inhabit the marſhes, eſpecially at the mouths of the Koiſu and Kurr. In the barren and ſaline parts of Schirvan, particularly between Baku and Sallian, and between this and New Schamachi, we meet herds of gazels. They are the *antilope ſubgutturoſa* of Gueldenſtædt, are ſomewhat ſmaller than the common goat, and of a paler colour, eſpecially on the ſides, round which runs a pale ſtripe rather indiſtinct. The crop is not very viſible. The tail is about five inches long, and terminates in a tuft of black hair. Both ſexes have ſingle black horns, placed in the ſame direction, and nearly of equal ſize with thoſe of the houſe-goat. They are compoſed of from twelve to eighteen raiſed volutes from the roots upwards; at the points, on the contrary, which are turned inwards and project a little, they are ſmooth and poliſhed. The inhabitants call this animal, as well as the

the goat, *tſchairan;* its fleſh is eaten, but is no particular delicacy.

The hare, of the ſpecies that does not change colour in winter, is very common, particularly in the woody parts of the plain. We find, likewiſe, badgers, pole-cats, weaſels, and hedge-hogs. Lovers of zoology may obſerve here various kinds of ſmall animals of the gnawing tribe; but the common houſe rat is not indigenous. The jerboa, a middle ſpecies between the mouſe and hare, is diſtinguiſhed by its particular form. There is a deſcription and print of it in Gmelin's Travels, but the latter is not very correct, as it repreſents the animal in a ſitting poſture much too extended. The jerboa inhabits the moſt barren parts in the vicinity of the naphtha wells.

Of the claſs of birds, I have ſeen very few that have not been deſcribed by Gmelin, who extended this part of natural hiſtory more than the reſt. Yet I do not doubt that, by time and opportunity, ſomething more might be diſcovered. We meet with eagles of extraordinary ſize, of the vulture-tribe peculiar to hot climates. The noble falcon is alſo indigenous, and much prized and uſed by the Khans and great men

in

in hunting. Birds of prey are generally in abundance, as they find plenty of food amongſt the ſmall quadrupeds and birds. There is an uncommon number of large and ſmall aquatic birds in the neighbourhood of the ſea, eſpecially at the mouths of rivers; ſuch as ſwans, wild geeſe, various kinds of ducks, cormorants, water-hens, bitterns, moor-hens, ſtorks, cranes, herons, and different ſorts of lapwings and mews. My ſhort ſtay on the coaſt prevented me from making more accurate obſervations on theſe ſpecies; I muſt, therefore, refer entirely to Gmelin. Different ſorts of ſnipes, and other birds of the ſame genus, abound in marſhy places, even in the interior. I ſhould imagine, that an attentive obſerver, who had time to devote to ornithology, would diſcover many ſpecies quite new, or at leaſt very little known.

The pheaſant lives here in the country aſſigned to him by naturaliſts, by the denomination of the Colchian, or Caucaſian pheaſant *(phaſianus Colchicus.)* In fact, they are almoſt as common in the woods and among the ruſhes on the Kurr, as the grey and red partridge. Of the latter there are two kinds, which we never meet with in

Ruſſia,

Ruſſia, and generally belong to the rarer ſpecies; *viz.* the Pyrenæan, (*tetrao alchata,*) and the Indian (*tetrao francolina,* Lin.) or francolin of Tournefort. Gmelin has deſcribed the former as the *tetrao caudatus.* They particularly frequent the Kurr and Sallian, and in winter flock together in great numbers, eſpecially when the ſevere weather ſets in.

Gmelin, merely from the name, has mentioned the francolin as indigenous to Gilan; it principally inhabits low places covered with reeds along the Kurr, and is by no means rare. With regard to its form and manner of living, it approaches neareſt to the common partridge, but is ſomewhat larger. As there are many apparent contradictions amongſt ornithologiſts in regard to this bird, the following more minute deſcription may not be ſuperfluous:

Statura perdicis, at paullò major. Rostrum aduncum gallinaceum, nigrum. Caput nigro-nitidum, vertice ruſo nigroque vario plumis scilicet in disco nigris, ad marginem rufescentibus; genis postice sub oculis maculâ albâ nigro-punctatâ conflatâ e plumulis subtilissimè setosis, rigidis erectisque. Collare latum ferruginosum. Pectus et abdominis basis nigro-nitida maculis rotundatis candidis, qui color unàcum maculis alarum quoque basin occupat, et sub collari in superiorem colli partem angustè excurrit.

excurrit. Maculæ crebriores ad collare et alarum basin, versus medium pectus rariores sunt, et nonullæ laterales lunatæ. Dorsum et alæ infra medium nigro-rufo pallidoque varia: singulis scilicet plumulis in disco nigris utrinque lineâ exalbidâ margine parallelâ, ipso autem margine rufo, cauda brevis, apicem versus aliquantum attenuata, nigra, lineis transversalibus parallelis, undulatis. Abdomen anum versus albidum, maculis sparsis ferrugineis et nigris. Femora albo nigroque varia. Anus collari concolor. Remiges nigræ macularum magnarum, orbicularium ruforum duplici serie. Rectrices basi fasciis albis nigrisque apice utrinque et subtus totæ nigræ. Pedes nudi, rufi, tetradactyli, intermedio longiore, unguibus nigris; tubore obtuso loco calcaris. Femina paullo minor, collari obsoleto; capitis pectoris baseos alarum et abdominis colore maculisque pallidioribus et in rufum vergentibus.

The black cock, and cock of the wood, are never ſeen in thoſe countries. Buſtards appear but ſeldom; but many thouſands of the *otis tetrax* come from the northern parts towards winter. Quails, pigeons, ſtarlings, and many kinds of larks, particularly the large ſpecies with the black ruff (*alauda calantica*, Lin.) are in abundance. I muſt paſs over thoſe birds of the claſs of peckers, ravens, and ſparrows, as I had very little opportunity of making accurate obſervations on them.

Birds

Birds of paſſage paſs in immenſe large flights after the end of October; the major part do not ſettle, but ſeek the large rivers of Perſia and Turkeſtan in autumn, and thoſe of Ruſſia in the ſpring.

Of the amphibious claſs, I have commonly met with the river-tortoiſe (*teſtudo lutaria,* Lin.;) and am convinced that the ſpecies deſcribed by Gmelin (*teſtudo Caſpica*) as a new one, is the ſame with ours. During nearly a three month's reſidence by the little river Pirſaget, and on the very ſpot where he pretends to have firſt obſerved the Caſpian tortoiſe, I found the common river tortoiſe in abundance, but never one of a different ſpecies. Beſides, his deſcription perfectly agrees with the former, except that the fore-feet are mentioned as having five toes. I muſt, therefore, conſider it as an error in obſervation. We not only meet with theſe tortoiſes in water, but in more elevated ſituations; and I have ſeen them even in the ſalt plain near the Kurr, where they probably live on ſnakes. All kinds of lizards are found principally in dry ſtony places. I ſhall only mention the *lacerta apoda,* which is not rare in the plain of the Kubaic Domain. All ſorts of ſnakes, but none that

are

are unknown in Europe, are in abundance; yet not more ſo than in many parts of the Ruſſian empire, particularly the Taurid. On the contrary it is maintained, that they are almoſt innumerable on the right banks of the Kurr and the Aras, in the Moghanic deſert, as well as in the Khanſhip of Talischumi. The latter has given riſe to various ſuperſtitious tales amongſt the natives; and even the moſt enlightened confidently aſſert, that the number of ſnakes is ſo great as to render it impoſſible to march an army over the plain of Moghan during the ſummer months. I am of opinion, that the want of green fodder, and of water, would be far greater impediments than the poiſon of ſerpents; particularly, as a conſiderable part, and preciſely thoſe which are moſt dreaded on account of their ſize, are not venomous. But it is worthy of remark, that this prejudice has reigned two thouſand years; for, we read in Plutarch that Pompey was prevented from marching along the Kurr down to the Caſpian, owing to the number of ſerpents.

I have nothing remarkable to obſerve concerning the inhabitants of the waters; and ſhould think that not much novelty was to be expected in that branch. The ſmall rivers are too rapid to contain fiſh; and the large ones, *ex. g.* the Kurr, &c. are the ſame, except at their mouths. The fiſh in the Caſpian belong to the Ruſſian ſpecies, which have been long known.

The hiſtory of inſects certainly promiſes conſiderable advantages from the labours of entomologiſts. They might not be derived from the admired claſs of butterflies; but the learned would be remunerated by the qualities of others, particularly of the ſpecies of graſshoppers, locuſts, and crickets, (*mantis*, Lin.) The turpentine-piſtachio tree feeds a great number of large locuſts, remarkable for their ſhrill tones in ſummer. Scorpion-ticks, tarantulas, and ſcorpions, would excite the attention of the traveller, though in a far more unpleaſant manner. We find them underground in dry places, and amongſt looſe ſtones. They, therefore, abound moſt in the plain between the Rubas and Ataſchai. The inhabitants, as well as the Grebenſki Koſaks, ſay that they are found in thoſe parts covered with the dwarf wormwood; this agrees with our obſervations, for that plant grows commonly in dry, heavy ſoils, and amongſt ſtones.

Of

Of the three ſpecies of inſects, the ſcorpion-ſpider *(phalangium araneoides)* is doubtleſs the moſt dangerous. It is three inches in length, is perfectly ſimilar in form to a large ſpider, and diſtinguiſhed by the length of its body, and the colour, which is between pale grey and a clear brown. The back is marked with a long black ſtreak; the legs are long and thin, and covered with hairs like the body. The upper joint of both hind feet has the property of being caſed with ſix tranſparent ſcales, erect and pediculated. To the head are attached two poiſonous bladders as large as the head, each being furniſhed with a pointed forceps placed vertically, which ſerve for biting; whilſt the animal, at the ſame time, draws ſome of the poiſon into its mouth. I have ſeen different inſtances of the effects of the poiſon, which convince me of the truth of the obſervation, that it is the moſt deadly of all animal poiſons, except, perhaps, that of the rattleſnake. In a few moments inflammation takes place throughout the body, with all the ſymptoms of the blood being violently heated, which inevitably end in death, if the common remedies be not applied.

The real tarantula is much leſs dangerous; and we may believe, that moſt of what has been related formerly concerning the bite of taranttilas, ſhould be attributed to the ſcorpion-ſpider. However, I muſt confeſs, that the tarantula is much more rare in thoſe countries than the ſcorpion-ſpider; and beſides, never ſo large as thoſe which I have ſeen in Moldavia, and in various parts of ſouthern Ruſſia. The tarantula has nearly the form of the large houſe-ſpider. It is of an aſh-grey colour, the back marked with black ſpots and lines; the belly, on the contrary, coal-black and velvety. The legs are of the ſame colour with the body, and black ringſtreaked.

The ſcorpions are of that ſpecies which are found in great numbers in all parts of ſouthern Europe, and therefore called by naturaliſts the European. Their bite occaſions a local inflammation, that ſcarcely exceeds the ſting of a large hornet; conſequently cannot be reckoned very dangerous.

AGRICULTURE.

Although conſtant civil wars, and the reſtleſs ſpirit of the inhabitants, have been highly

highly prejudicial to agriculture; ſtill their neceſſities have not allowed them totally to neglect this branch of rural œconomy. For, their ſituation and intercourſe with their neighbours, and particularly their inimical ſentiments towards ſtrangers, have always impeded, or prevented entirely, importation from other countries. They grow a ſufficient quantity of corn for themſelves and their cattle. The plain of Dagheſtan near the ſea, ſome diſtricts at the foot of the mountains in the province of Kumuk and near New Schamachi, but eſpecially the plain in the Kubaic Domain, are the beſt cultivated; and the latter can ſpare a little to their neighbours. Agriculture is alſo carried on in the higher mountains, and even the ſteepeſt acclivities are ploughed, where I have often ſeen the fineſt crops. The ſoil is certainly, upon the whole, heavy and cold, and we ſeldom find any black mould: yet the harveſts are plentiful; which is partly to be attributed to the quantity of land and change of crops, and partly to the circumſtance of frequent ſhowers, even in the hotteſt months.

Wheat is ſown for a winter crop, and barley for ſummer. Rye and oats are ſcarcely even ſeen; becauſe the natives are not accuſtomed to rye-bread, and their horſes prefer barley. The preparation for wheat is nearly the ſame as amongſt the Crim Tatars, and the plough likewiſe. On the mountains, harveſt commences in the middle of June. They uſe the common ſickle; but the ſtubble is left two or three feet high, probably becauſe the reapers do not like the trouble of ſtooping. The corn is threſhed on open floors, with the machine deſcribed by Gmelin. Millet is alſo grown, but moſtly in Dagheſtan; the natives being ſo partial to the ſpirit extracted from it, which turns ſour, and is called *buſa*. It is frequently made into bread; and in thoſe countries where the vine is not cultivated, a vinous liquor is diſtilled from it. Except in the plain between the Rubas and the Ata, rice is almoſt unknown; there it coſts very little labour to let in the water by means of canals, and thus form a kind of artificial marſh, ſo neceſſary to the production of rice. The quantity grown is very diſproportionate to the whole conſumption of the country; but a great deal is imported from Gilan, at a moderate price, for the uſe of Baku and Schamachi. Lentils are grown in tolerable abundance to mix with *pilau*, the favourite ori-

ental

ental diſh. The *latyrus ſativus* and *cicer arietinum* are ſown here and there. The latter, botanically deſcribed by Gmelin under the name of Nachot, is commonly made into a preſerve like almonds. Common peas are very ſcarce. Turkiſh wheat, field-beans, topinambours, and potatoes, are entirely unknown. Near the cities, particularly New Schamachi, they cultivate a little ſeſamum and cotton. Hemp and flax are not grown. Melons, cucumbers, water-melons, and gourds, we find of all kinds, and of the beſt quality. The former laſt from the end of June to the cloſe of autumn. The ſmooth-ſkinned ſort with green pulp is particularly delicious, and in high requeſt, being ſcarce. I have ſeen them at the Khan of Schamachi's table, who treated his gueſts very ſparingly with them, as a *bonne bouche.* He mentioned that Nadir Shah firſt introduced them from India, and often afterwards ſent them as preſents to grandees. The culture of ſaffron belongs ſolely to the inhabitants of Derbend and Baku. The former, eſpecially, furniſhes ſufficient to ſupply the whole country, and even for exportation. Almoſt all the ſides of the hills round the city are covered with it.

 The

The Armenians only grow vegetables, which are confined to the commoneſt ſorts; ſuch as cabbages, turnips, radiſhes, parſnips, ſpinage, lettuce, French-beans, parſley, and different kinds of onions, leeks, and garlick. The poppy is plentiful, as the ſeed is ſprinkled over all kinds of paſtry, even the commoneſt white bread. We meet, alſo, with capſicums, ſo much prized in the eaſt, and a ſpecies of nightſhade (*ſolanum melongena*) that bears a fruit of the ſize of a cucumber, of a long form, and of a violet colour. It makes a delicate diſh called *pottitſchan*.

The culture of the vine is in the hands of the Armenians, and carried on principally in the plain between the Rubas and the Ata, around Derbend, and on the mountains between Old and New Schamachi, near the Armenian villages. The Schamachi wine is moſt eſteemed, probably becauſe the vines grow on the ſouthern acclivities. Upon the whole, the management is particularly bad; I might even ſay, that every fault which could be committed is put in practice. In the firſt place, the want of hands is the cauſe that the ſoil is neither dug deep enough nor ſufficiently often, nor is proper attention paid to weeding. The vines ſtand too thick,

are

are not well trained, nor often enough renewed. The choice of the beſt kinds is entirely neglected; hence we find bunches of grapes as large as prunes, and in great plenty, but they do not make good wine. The vintage commences almoſt a month too ſoon, for the natives, fearful every day in turbulent times of loſing the fruits of their labour, have contracted the habit of beginning too early even in peaceable ones. During my reſidence in 1796, I ſaw grapes in the Schamachi market in the middle of July, which naturally were unripe, for the vines do not bloom before the middle of June. The vintage was finiſhed about the 20th of September. For preſerving of wine, they have neither good cellars nor proper caſks. They uſe large earthen jars with narrow necks, frequently of more than two feet diameter, and are kept under ground. When it is to be removed, it is put into leathern bottles ſteeped in naphtha. The conſequence of ſo much bad management is, that we ſeldom meet with wine even paſſable; and the expreſſion *tſchachuer*, meaning wine in the language of the country, is received amongſt us Europeans and uſed in contra-diſtinction, as if both liquors were totally different.

The

The *tschachuer* is moſtly red, has very little ſpirit, a rough taſte, and will not keep. The latter is of very little conſequence, for the whole vintage, which they begin to drink as early as October, is expended in a few months. If it is to be exported, it certainly keeps rather better in leathern bottles, but imbibes ſo much of the naphtha, that ſtrangers at firſt cannot drink it. From all I am perſuaded, that the ſituation and climate, eſpecially about Schamachi, are well adapted to the culture of vines, and that light wines might be made almoſt equal to French red wines.

Plantations of mulberries are no longer plentiful. The largeſt are in the Kubaic Domain, near New Schamachi, and here and there in Dagheſtan, particularly about Tarki, in the territory of the Schamgal. The trees ſtand very near together, and are kept ſo low that they appear like ſhrubs. We find ſome of the red and white ſorts in the vicinity of cities and villages.

Orchards produce apples, pears, quinces, cherries, prunes, apricots, peaches, almonds, granates, figs, and walnuts. It muſt be acknowledged, however, that the major part are inferior, both in variety and flavour, to

thoſe

thoſe in the north of France, in the ſouth of Germany, and even in ſome parts of Ruſſia. The beſt climate for the finer kinds of apples, pears, apricots, and peaches, appears to be that of countries under the temperate zone. It is well known, that they do not ſucceed ſo well, in general, in the moſt ſouthern parts of Europe; and this holds equally good in regard to the Tauric Cherſoneſe. This country does not produce oranges, as it lies to the northward of Caucaſus: we firſt meet with them in Gilan.

CATTLE.

The breeding of cattle is nearly upon the ſame footing with agriculture; conſequently, far removed from the perfection which it might be brought to. As important as it is for every perſon to be well mounted, owing to continual civil commotions, yet the breed of horſes is, upon the whole, equally neglected. We meet no where with large ſtuds, for thoſe who trade in war, and live by plundering others, as well as the opulent Armenian and Georgian merchants, procure horſes from Leſgiſtan or the ſouth of Perſia. The Kumucks and Dagheſtaners are better

provided

provided with horſes than the natives of Schirvan, who moſtly ride aſſes and oxen, and frequently buffaloes. The indigenous race appears to hold a middle rank between the Circaſſian and Turkiſh-Anatolian. They are of middling height, with beautiful heads, vivid eyes, and wide noſtrils with uncommonly thick edges. They are ſlim and thin-legged, docile, good gallopers, but not capable of hard-work, particularly if not well and regularly fed; for the high price of horſes is the reaſon why they are accuſtomed to the beſt food from their youth. They are ſeldom turned out to graſs. Even in the ſtable they are ſcarcely ever fed on hay or graſs; their common fodder is the ſhort ſtraw, after threſhing, mixed with all kinds of ſeed and refuſe corn. They have two feeds a day, beſides, of barley previouſly ſteeped in water; and look extremely well with this kind of food. The attention paid to the cleanlineſs of horſes is extraordinary. Perſons of rank are not aſhamed of daily cleaning their own horſes. The gelding, as generally throughout the eaſt, is not common; ſtallions being preferred for riding. To prevent them from injuring each other by kicking in the ſtalls, they have a clever

method

method of tethering their hind legs to a ſtake at ſome diſtance behind the animal. This anſwers a double purpoſe; namely, more room is gained by having no ſtalls, and the horſes become accuſtomed to ſtanding confined.

The aſs is as general as it is an uſeful animal, eſpecially in the mountains. Mules are ſold cheap in the country on the right bank of the Kurr. The cattle are ſmall, but ſtrong, and moſtly dark brown or black; and preciſely ſimilar to thoſe in the Taurid. They are employed as beaſts of burden and for riding; the Tatarian carts, called *arba*, being very ſeldom ſeen in Schirvan. It is impoſſible to conceive a kind of *arba* ſimilar to thoſe in the neighbourhood of Baku or Sallian. The wheels are of eight feet diameter, fixed to the axle-tree, which revolves round two ſtaples underneath. The Nogays, who inhabit the plain of Kumuk, poſſeſs the largeſt herds of cattle. In all the diſtricts containing marſhes, we find buffaloes; they live there in the hotteſt ſeaſons, and thrive beſt in ſuch ſituations: conſequently, they are more plentiful in the plain between the the Rubas and the Ata, than other cattle. We ſeldom ſee the camel with two humps in in the vicinity of Baku; although that barren ſoil, overgrown with marine plants, offers the beſt ſituation for breeding them.

The provinces are uncommonly favourable for breeding of ſheep, as they always find good feed in ſummer on the mountains, and in the plains during winter. But the flocks are not numerous, and are kept more for conſumption than for any advantage from their wool. They are the Tatarian-Kalmuck breed, with fat tails. The meat is delicious, but the wool very poor. In all low diſtricts near the ſea, large droves of ſheep and goats are fed during the winter. They moſtly belong to the Leſghaes, who, on account of the ſnow on the mountains, are obliged to ſeek the plains with their herds.

Observations on the Inhabiiants, their Manners, and Form of Government.

All the obſervations made by Gmelin on the inhabitants of Gilan; on the phyſical and moral character, manners, religion, and conſtitution of the Perſians; may be applied to the inhabitants of theſe provinces with a ſlight

ſlight modification. But, in my opinion, he has carried them too far, having principally extended them to the Perſians; conſequently to a very ſmall part which he had ſeen, and the remoteſt from the ſeat of monarchy, as they live to the northward. I am, therefore, exonerated from any detailed deſcription, and ſhall content myſelf with ſome fugitive remarks by way of concluſion.

The difference, between theſe people and the real Perſians, proceeds from their mixture with the Caucaſian nations, eſpecially with the Tatars who live N. W. of the Caſpian. I am even perſuaded, that they deſerve the name of Caucaſian Tatars much more than that of Perſians; although the latter is applied to the natives of Schirvan and Dagheſtan, owing to their irregular alliances with the new monarchy. Their language and religion ſerve to confirm my opinion. The former is a dialect of the Tatarian, and does not differ more from that of the Taurid, than the language of Great Ruſſia does from Little Ruſſian. The *Parſi* is entirely unknown, and only uſed in public acts, now and then, by the Schirvanic Khans whoſe families came from the interior of Perſia. Mahommedaniſm is the predominant

nant religion; but the doctrine of *Ali,* that has prevailed some centuries in Persia, is spread only in the cities of Schirvan, where it was introduced by the Khans and their courts. The Kumuks, Daghestaners, and most of the Schirvaners, as well as the remaining Caucasian Mahommedans, with the Tatars, (their neighbours to the N. W.) are attached to the sect of Omar.

They do not equal in beauty the Tzirkassians, (*Tscherkess,* properly pronounced,) or Georgians; yet are not devoid of strength and comeliness. We observe no national physiognomy, as amongst the Tzirkassians, Georgians, Armenians, and Tatars; but a general mixture has produced features of which little can be said upon the whole. The men are of middling stature, thin, dark-coloured, and have black hair. The women are far inferior to the beautiful Georgians and Tzirkassians, and the harems are mostly composed of the latter; but if they do contain beauties of the country, they are generally native Armenians; for, throughout the east, Armenians only marry amongst themselves, like the Jews.

The national moral character is not superior to the physical one. A spirit of rapine,

pine, ſloth, pride, and contempt for every thing foreign, are united to craftineſs and baſe ſervility; an inceſſant bravado of power, with the intrigues of the weak and defenceleſs, are the characteriſtic traits, and barbariſm and extreme miſery their melancholy conſequences. The Dagheſtaners in Perſia are reckoned good troops; and the Aguans in Kandahar, ſo famed for their valour, are ſuppoſed to be deſcendants from them: notwithſtanding, the natives of theſe provinces cannot withſtand European tactics; for our artillery conſtantly threw them into a panic, and cauſed them to give way at the firſt diſcharge.

I ſhall paſs over their religious ceremonies, marriages, burials, &c. as it would only be a repetition of thoſe of other Mahommedan nations ſo often deſcribed.

In general, they are moderate livers. They take two ſimple meals a day, and the major part are tolerably abſtemious in regard to wine and ſtrong drink. The rich and great in the cities, particularly in Dagheſtan, certainly ſet themſelves above the Koran in this point; but the people ſeldom have an opportunity of procuring wine and brandy, as the Armenians only keep a ſmall ſtock.

 The

The common wheaten bread is baked in the form of thin cakes, and called *tſchurek.* They are moſtly very white, and well-flavoured, eſpecially thoſe ſold in the cities. Derbend is famous for its good bread, and Buinaki (a ſmall city in the territory of the Schamghal) for its cheap bread. In Dagheſtan, particularly in Tarki, a great deal is made of millet; it is heavy, but uncommonly ſatiating. Their favourite diſh is *pilau*, or *plov.* The rich prepare it in different ways, and colour it with ſaffron and other harmleſs plants, as a luxury. Thoſe who do not grow rice, ſupply its place by millet-grits. Milk, butter, cheeſe, and fruits, conſtitute their chief food. There is a variety of other diſhes at the tables of the regents and great men, yet they moſtly conſiſt of rice and mutton. Whoever is accuſtomed to European cookery, will find them too greaſy, and covered with all kinds of ſpices, more eſpecially ſaffron. Vine-leaves, cucumbers, water-melons, onions, and garlick, are alſo uſed. A fine kind of paſtry, with almonds, ſugar, and preſerved fruits, is always ſet on the table. However, upon the whole, the national grand dinners do not equal the European.

The

The Tatarian beverage, of ſour milk diluted with water, called *airan*, is in common uſe. It is cooling, and tolerably palatable when mixed with a little fennel and freſh cucumber. They have alſo different ſorts of ſherbet, made either of honey or of the juice of fruits: that made of pomegranate is particularly eſteemed. Tea and coffee are ſeldom uſed; but every one ſmoaks, not excepting the women. The tobacco is uncommonly ſtrong, and muſt therefore be ſmoked in a *kallian*, which is deſcribed in Gmelin; but, if it is to be uſed in common pipes, it muſt be previouſly ſteeped in water.

The coſtume differs from the Tatarian and Caucaſian, principally in the cap and the dreſs of the feet. Inſtead of the hemiſpherical cap of the Tatars and Tzirkaſſians, which is of a terrible length, they wear high, broad, felt-caps, with a border of ſheepſkin turned up; but the Kumuks uſe the Tzirkaſſian dreſs. The Khans and great men have their caps made of finer ſkins. The dreſs of the feet is nothing near ſo light and pleaſing to the eye, as amongſt the Tzirkaſſians; even Morocco-boots are ſeldom worn, as among the Tatars. The feet are wrapped up in cloth or ſtuff and faſtened with a band; so that this part of the costume resembles the Russian peasantry. The shoes are of Morocco, with narrow heels like women's. On horseback the covering of the feet is tucked under, or drawn over their wide pantaloons. In short there is very little luxury in dress, even amongst the great. The shirt, the waistcoat with arms, and pantaloons, are commonly made of woollen, and the upper loose robe of cloth. The Khans and great men wear finer woollens and India muslins; and their *kaftans* are of Dutch cloth, which the Armenian merchants import from Russia. Instead of a girdle, the rich wear costly shawls, and wind them also round their caps. The word *shawl* includes every kind of woollen, manufactured in the country or imported from other parts of the east. The fine sort, at present become an article of luxury with our European *belles*, is distinguished by the name of Persian and Cashmire shawls. The others are inferior and cheaper, and commonly sold in the shops. The rich, in the neighbourhood of Schamachi, have their under-dress of silk manufactured there. The ordinary *kaftans* are of a coarse half-cloth made by the women.

The

The female coſtume in Schirvan differs very little from the Tatarian, yet the pantaloons are larger, and the robe ſhorter. Moſt of the Dagheſtanic and Kumuk women dreſs entirely like the Tatarians and Tzirkaſſians. Neverthеleſs, we may eaſily imagine, that the fair ſex have not been ſo ready to renounce perſonal decoration as the men. Hence, gold-ſtuffs, embroidery, lace, necklaces, bracelets, and girdles with gold and ſilver claſps, are not baniſhed from the harems of thoſe who have property and inclination to favour the wiſhes and deſires of their fair priſoners. Both ſexes are accuſtomed to colour their hair, eye-brows, and nails, with red; beſides this, the women uſe a great deal of white and red.

The ſyſtem of conſtant war and plundering has compelled the inhabitants to go continually armed. The huſbandman ploughs, ſows, and reaps, with a dagger at his ſide, and has his gun at hand beſides. We never meet a traveller who is not armed from head to foot. The moſt common arms are guns, piſtols, ſabres, and daggers, or *kantſchals*. They moſtly come from Leſghiſtan, where ſome of the principal cities ſcarcely manufacture any thing elſe. They are of tolerably good materials; yet the rich and great, particularly the military, procure arms of a ſuperior quality from Koraſſan, which are often ſold at an enormous price. The dagger is the weapon generally preferred, and no perſon is completely dreſſed without one. It is rather more than twelve inches long, and about two in breadth towards the handle. It is double-edged, ſharp-pointed, and fluted on both ſides. The blade is generally oiled with naphtha, to prevent the ruſt. Arms of all kinds are very plain: the neceſſity of always wearing them, cauſes the natives to attend more to their quality than to ornament.

The breaſt-plate, ſo much eſteemed by the Tzirkaſſians, is not unknown here. The life-guards of the Khans and other regents, are provided with them. We often ſee, likewiſe, a kind of open helmet with little bells attached to it, perhaps to attract notice at a diſtance by ſome bravado.

As coſtlineſs is baniſhed from dreſs, it is the more uſed in horſe-furniture. The bridles and ſaddles of the Khans and great men are plated with gold and ſilver, and the houſings are very elegant. The ſaddle differs from the Tatarian and Tzirkaſſian, being

ſmall,

ſmall, light, and with a long projecting point inſtead of a pommel. The houſing is not laid under or over the ſaddle, but fixed to it behind, and covers the whole rump. The ſtirrups are long and broad, after the Turkiſh faſhion. The uſe of ſpurs is entirely unknown; but they are partly ſupplied by the ſtirrups, and partly by a ſhort whip with a double thong.

The miſery of the country ſhows itſelf particularly in the wretched ſtate of the public and private buildings. The devaſtation to which they are daily expoſed, naturally diveſts every one of the wiſh of having any ſubſtantial habitation, that they cannot enjoy, and would be afraid of loſing every moment.

The houſes are built of rough ſtones, with flat roofs covered with a coating of mortar. In Baku the lime employed for the roofs is tempered with petroleum, which is very ſerviceable for that purpoſe, as it keeps out the wet thoroughly. They commonly project in front over the principal entrance, and being ſupported by timbers, form a balcony, where the inhabitants ſit to enjoy the cool air. The rich always have one wing for their harem, which no one is allowed to enter. Every houſe in the cities has a court ſurrounded with a high wall, which the doors and windows look into; ſo that we ſee nothing in the ſtreets but blank walls, with low doors always kept ſhut. The interior contains ſome low rooms without any ornament, and the windows are made of lattice-work. The furniture conſiſts of carpets and cuſhions. The firſt are articles of luxury, and thoſe from Koraſſan are bought at a high price.

In cities, alſo, we meet with houſes two ſtories high; but the ground-floor is low, and an outward ſtair-caſe leads up to the balcony. Thoſe of the Khans and the other regents are in the ſame taſte, but diſtinguiſhed by their ſize and larger rooms, which are better furniſhed, in the oriental ſtyle. The moſks *(metſchett)* claim more attention. Their architecture is very ſimple, with circular roofs. But this ſimplicity, combined with ſolidity, gives them a pleaſing appearance of grandeur. They have one or more *minarehs*, from which the *muezzin* calls the hour of prayer. The caravanſerais conſiſt of a ſpacious court, moſtly quadrangular, with a gallery and chambers. Thoſe we meet with on the high roads are in a very

different

different ſtyle. They are built of large ſquare ſtones, with flat roofs and arched rooms, and round towers at the ſides: ſo that the caravans not only find ſhelter, but refuge alſo, where, in caſes of neceſſity, they may defend themſelves againſt banditti. The handſomeſt and moſt convenient are on the road from Mount Peſchparmak to Sallian, where there is no water but what the wells contain.

The perpetual revolutions, to which the peaſantry and the peaceable citizens are expoſed, have almoſt totally baniſhed ſcience. Beſides the common trades of taylor, ſhoemaker, ſaddler, ſmith, potter, &c. there are a few manufacturers, ſuch as ſword-cutlers, lace-makers, weavers, and Morocco-dyers. New Schamachi furniſhes, in particular, ſome ſilks, though of very middling quality. All the tradeſmen work, according to the eaſtern cuſtom, in open ſhops, which, with thoſe of different merchants, mercers, druggiſts, victuallers, &c. generally conſtitute the main ſtreets, and extend from one end to the other.

Notwithſtanding what is commonly ſaid of oriental deſpotiſm, and the unlimited

power

power of the regents in the interior of Perſia, yet the petty rulers of this country do not govern without controul. The Kumüks and Dagheſtaners, eſpecially, partake of the independent ſpirit of their neighbours the Leſghaes, and are ſubject only to their princes, in as far as theſe make themſelves beloved, and as thoſe find their advantage in obedience. The Khans in Schirvan have certainly more influence over their ſubjects; yet they muſt be very guarded how they undertake any important affairs without the advice of the Divan, which conſiſts of the chief Beys and military men. It is alſo conſulted in caſes of right, and particularly in criminal ones. Should the Khan, even once, pretend to condemn and puniſh as ſupreme, he has at leaſt to dread the diſpleaſure of his ſubjects. And that would be undoubtedly attended with fatal conſequences, if he were diſpoſed to uſe thoſe arbitrary meaſures which cuſtom rejects, although other written laws were in force, beſides thoſe of the Koran.

However, the limited power of the Khans does not, in the ſlighteſt degree, prevent the oppreſſion and miſery under which the major

jor part of their ſubjects groan. The nobility and military are certainly jealous of their influence in political and judicial affairs, but they too lightly concur with the views of the regents, when contributions are to be extorted from the Jews, Armenians, and other Chriſtians; or to annoy the neighbouring little ſtates by predatory incurſions, in which the nobility ſhare the danger and the ſpoils with their leader. Their non-allegiance to Perſia, which theſe lords have maintained for ſome time, inſtead of leſſening civil diſcord in the provinces, has increaſed it conſiderably; becauſe the complete impunity of the great is now combined with tyranny. The vicinity of the rude Leſghaes is a ſcourge to the country in a two-fold point of view; on account of their conſtant depredations, and alſo of their readineſs to ſupport any daring adventurer, who has accumulated a ſufficient ſum of money to ſubſidize as many of thoſe plunderers as he wants to conquer any part of the country, and to depoſe the regent, who, perhaps, had uſurped his ſeat by means equally unjuſt. In a word, few people are in ſo lamentable a ſtate; and it is heartily to be wiſhed that they were

able

able to enjoy peace and happineſs under a wiſe government, capable of protecting them againſt the barbarities of their neighbours and fellow-citizens.

APPENDIX.

Caracteres plantarum nunc primum detectarum, vel prius haud sufficienter observatarum et descriptarum.

I. SALICORNIA Caspica.

S. fruticofa, amentis cylindricis, fquamis bafi apiceque folutis obtufiffimis, flofculis ternis tryphyllis.

Salicornia arborefcens fine geniculis. Buxb. Cent. 1. pag. 6. tab. 20. fig. 1.

Salicornia Cafpica. Pallas Itin. 1. pag. 480 n. 90. Tab. A. fig. 2.

Frutex ramofiffimus, a 2 ad 10 pedes altus.

Caules lignefcentes, cortice cinereo læviufculo, bafi nonnihil declinati et ftatim adfcendentes.

Rami adultiores fubftantia cauli fimiles, iuniores articulati pro more generis. Articuli pro variâ ætate, variæ admodum formæ. Vernali tempore nondum florente frutice Salicorniæ herbaceæ fimillimi; in ramis floriferis ferè cylindrici fcilicet bafi apiceque perparum attenuati; denique in ramulis qui, peractâ fructificatione et maturefcentibus jam feminibus, paffim proveniunt, minuti, breves et fubglobofi.

Amenta perfectè cylindrica ad apices ramorum et ramulorum oppofitorum.

Amemti fquamæ duarum oppofitionum, bafi apiceque folutæ, carnofæ, obtufæ et præfertim poftice quafi truncatæ, provectâ ætate deciduæ.

Flofculi inter fingulas fquamas terni ab earum fubftantiâ feparati; quorum intermedius major in amento florente compacto magis confpicuus eft.

Calyx proprius tryphyllus, claufus; foliola duo majora, fornicata, fucculenta; tertium inferius carinatum, obtufum, minimum, ita quidem, ut calyx diphyllus videatur.

Stamina

Stamina haud vidi.

Semen calyce tectum, folitarium, fubglobofum, minutum, atrum.

Occurrit copiofè in planitiebus ad Cyrum fluvium nec non alibi paffim in locis falfugineis ad littora maris Cafpici ♄

Obf. Polymorphia articulorum in hâc fpecie probat, Salicornias male fecundum articulorum formam definiri: potius ad florum fitum et ftructuram recurrendum eft.

Synonymon Buxbaumii certiffimè ad noftram fpeciem fpectat, quippe de plantis ad Cyrum fluvium, in ipfâ, de quâ nos quoque loquimur regione lectis, fermo eft.

II. VERONICA PEDUNCULARIS. Nova.

V. Racemis lateralibus oppofitis; foliis ovatis incifo-ferratis, incifuris bafi fubdentatis; pedicellis filiformibus, calycybus quadrifidis fubæqualibus.

Veronica Chamædryos foliis parvis. Buxb. Cent. 1. p. 26. tab. 41. fig. 1.

Planta perennis, V. Chamædri affinis.

Caules è radice nonnulli, bafi vix declinati, dodrantales, fimplices, teretes, undique villofi, villorum tamen denfiorum lineâ utrinque è petiolorum bafi laterali, ad infequentium foliorum axillas excurrente.

Folia oppofita, breviter petiolata, cordato-ovata, incifo-ferrata; incifuris anguftis non acuminatis, fæpe denticulo fupra bafin lateralim inftructis; cæterum utrinque villofa, villis marginis longioribus. Folia fuperiora confimilia, nifi quod, paullo longiora et anguftiora.

Racemi laterales oppofiti, plerumque tantum bini, pedunculati, laxi. Pedicelli tenues, elongati, patentes. Bracteæ minutæ, oblongæ, integræ.

Flores Chamædri minores.

Calyx quadrifidus, nullo rudimento folioli quinti; laciniis fubæquallibus, oblongis, fed non acuminatis, ætate latefcenticies, longitudine capfularum.

Corolla, Stamina, piftillum ut in V. Chamædri.

Capfula

Capſula lævis, obcordata, loculis ſubtriſpermis.

Pedunculi, pedicelli, bracteæ, calyceſque pubeſcunt.

Lecta rarius in umbroſis ſylvaticis circa Kurt-Bulak. Floret Majo; Iunio ſemina perficit. ♃

Obſ. Differt à Veronicâ Chamædri cui proxima; caule undique villoſo, quamvis utrinque lima villorum denſiorum exiſtat; foliis utrinque magis villoſis, profundius inciſis, inciſuris ſæpius baſi poſtica denticulo inſtructis; pedicellis magis elongatis; denique calycis laciniis obtuſioribus.

III. ZIZIPHORA SERPYLLACEA. Nova.

Z. ſuffruticoſa, racemis terminalibus capitatis, foliis ovato-lanceolatis ſubſerratis, floralibus conſimilibus.

Suffrutex humilis, habitu Serpylli.

Radix lignoſa, valida.

Cauliculi fruticuloſi, declinati, pubeſcentes; rami floriferi adſcendentes.

Folia magnitudine foliorum T. Acini, ovato-lanceolata, acutiuſcula, ſæpe ſerratura unâ alterave obſoleta inſtructa, leviter pubeſcentia, margine non ciliatâ, ſubtus nervoſa et excavato punctulata.

Racemi ad ramulorum apices ſubcapitati, forma Serpylli.

Flores formâ et colore Ziziphoræ tenuioris.

Calyx linearis, 10-ſtriatus, pubeſcens; denticulis quinque parvis ſubæqualibus.

Corollæ tubus calycem vix excedens. Faux parum inflata. Labium ſuperius breve, obtuſum, apice leviſſimè crenulatum. Labium inferius trifidum; laciniis ſubæqualibus, lateralibus attamen paullo latioribus, intermedia reflexa vix emarginata. Lineolæ punctatæ ſaturatiores ad faucem.

Stamina 2, circiter longitudine labii ſuperioris conniventia, in quibuſdam floſculis non raro caſtrata, ſcilicet antheras effœtas gerentia.

Stylus adultior labio ſuperiore longior.

Stigma bifidum, acutum.

Planta odoratiſſima, odore omnino Ziziphoræ tenuioris.

Copioſa

Copiosa ad rivulorum margines lapidosos, et in collibis apricis regionis sylvaticæ inter Kubam et Veterem Schamachiam. Floret Junio, Julio ♄.

Obs. Proxima huic speciei est planta Sibirica, quam in systemate sub Ziziphorâ acinoide, deinde sub Cunilâ capitatâ bis occurrere suspicor. Priorem Linnæus pater jam introduxit, allegato synonymo Clinopodii supini incani: Ammanni Ruth 66; posteriorem filius additit in Supplemento. Utrique patria assignatur Sibiria; utraque Linnæis è siccis tantum speciminibus cognita fuisse videtur.: Atqui Amanni Synonymon apertè ad Cunilam capitatam trahi potest, dum perpaucos tantum flores infra caulis apicem prodire, plerosque vero in capitulum terminale collectos esse ipse Ammannus asserit; porro plentæ in horto Demidoviano, qui Mosquæ est dudum sub Ziziphoræ acinoidis nomine cultæ, quarum specimina coram habeo, nullos planè extra capitulum terminale flores gerunt; denique una tantum eademque species, nec duæ distinctæ et affines, in variis per Rutheniam hortis botanicis colitur è Sibirias transmissa; quibus ergo perpensis, vix dubito, quin vel Cunila capitata, vel Ziziphora acinoides ex recentiorum Systematibus sit eliminanda. Bene hanc speciem Sibiricam definit Cl. Lamark Tabl. Enc. Bot. n. 269. sub Ziziphorâ clinopodioide; sed Cunilam capitatam, autoritate Linnænâ adductus sub n. 273. conservat.

Species nostra Caucasica ab affine Sibiricâ sequentibus differt: Habitu minore magisque diffuso; foliis minoribus, angustioribus et acutioribus, subserratis; floralibus reliquis nec majoribus, nec insignius nervosis; floribus duplo minoribus staminibus brevioribus.

IV. CROCUS Speciosus. Nov.

C. spatha radicali uniflora, foliis tardissimis, staminibus stigmate tripartito ramoso brevioribus.

Crocus autumnalis sativo similis, florum capillamentis tenuissimis minus odorus. Tournef. Corr. p. 25.

Copiosissimus in collibus, agris, pascuis. Floret Septembri. ♃.

Obs.

Obſ. Quatuor mihi innotuerunt Croci ſpecies, quas omnes in Tauriâ ſponte creſcentes obſervavi; duas præcoces, duas autem ſerotinas: priores ſunt; Crocus vernus et luteus, poſteriores Crocus autumnalis et ſpecioſus. Omnium characteres et deſcriptiones Floræ Tauricæ, quam jam elaboratam prælo proximè committere propoſitum habeo, inſeram.

V. POA Pungens. Nov.

P. ſpicis compoſitis diſtichis, ſpiculis ovatis ſubtrifloris, culmo repentè.

In glareoſus ſubſalſis planitiei ad Cyrum inter oppida Schamachiam et Sallianum juxta viam paſſim integras plagas occupat; magnum in peregrinatorum ſolatium jumentis unico propemodum in aridiſſimo tractu pabulo inſerviens. Floret Majo, Iunio. ♃

Obſ. Deſcriptionem hujus ſpeciei, in Cherſoneſi Tauricaæ maritimis antea obſervatæ, Flora noſtra Taurica exhibebet. Spiculæ mihi tam in Tauriâ, quàm in Caſpicis ditionibus conſtanter trifloræ vel quadrifloræ viſæ, Poam eſſe nec Agroſtidem probant: reliquæ vero notæ omnes ſummam affinitatem cum Agroſtide pungente Schreberi et Lamarkii probant.

VI. HORDEUM Secalinum. Lamark.

H. floſculis omnibus ariſtatis glabris, lateralibus maſculis, foliis vaginiſque glabris, radice repentè.

Hordeum ſecalinum: floſculis lateralibus maſculis ariſtatis, involucris ſetaceis, ariſtus brevibus. Lamark Tabl. Enc. n. 1155. Flore Franc. 3 p. 623.

Gramen ſpicatum ſecalinum minus. Vaill. Paris. p. 83. tab. 17. fig. b; ſed folia juſto breviora.

Hordeum floſculis lateralibus maſculis ariſtatis, calycum valvis ſetaceis. Var. β. ſpica amœne purpurea. Gmel. Sib. 1 p. 125.

Lectum in graminoſis montanis fertilioribus, imprimis circa Kurt.—Bulak. Floret Iunio, Iulio ♃.

Obſ. Solum è recentioribus Cl. Lamarkii Synonymon procul dubio ad meam plantam ſpectat. Hordeum ſecalinum. Schreb. Spic. Fl. Lipſ. p. 148 diſcrepat foliis mollibus piloſis; porro H. ſecalinum Roth. Fl. Germ. III. 1 p. 150. radice annuâ, foliis mollibus piloſis, ſpicis brevioribus, floſculis lateralibus filiformibus ſtaminibus piſtilloque deſtitutis. Propius ad noſtrum accedit Ejuſdem Hordeum maritimum loc. cit.; differt tamen vaginis inferioribus hirſutis, floſculis lateralibus dorſo villoſiſſimis.

VII. HORDEUM Maritimum. Linn.

H. floſculis omnibus ariſtatis glabris, lateralibus ſterilibus, follis vaginiſque inferioribus molliter villoſis, radice fibroſâ annuâ.

Hordeum geniculatum culmo decumbente foliis obtecto, ſummo ſpathaceo ſpicæ ſubjecto, ſupremo nodo incurvato. Allion Pedem. II p. 259 n. 2274. Tab. 91. fig. 3.

In arenoſis ad mare Caſpicum, nec non ad rivulorum viarumque margine lapidoſos non infrequens. Floret cum præcedente. ⊙

Obſ. Graminis etiam in Tauriæ maritimis ſteſilibuſque copioſi deſcriptionem Floræ Tauricæ reſervavi. Ab affinibus Hordeo ſecalino et murino facilè dignoſcitur: altitudine vix ſpithamali, culmo ad genicula fracto, (unde nomen triviale Allionii) colore glauco, pube foliorum vaginarumque inferiorum molliſſima. Synonymon Allionii certè huc pertinet, quamvis folium ſummum ſpathaceum dici nequeat: culmus tantum ſolito tardius elongatur, adeoque ſpicam, poſt egreſſum, ab ultimo folio tardius removet. Cæterum Icon Allionii rudis admodum, vix niſi habitum plantæ exprimit.

VIII. SCABIOSA Columbaria.

S corollulis quinquefidis radiantibus, foliis radicalibus primigenis ſimplicibus crenatis baſi auriculatis, caulinis compoſitis, ſummis linearibus, feminum coronula anguſtiſſima, ſetis diſci longiſſimis.

Radix

Radix biennis, brevis, apice fibrosa, extus nigricans.

Caulis erectus, firmus, basi simplex, sæpe bipedalis, teretiusculus, infernè purpurascens, ad genicula albo-villosus, cæterum ferè glaber; supernè divisus in ramos elongatos floriferos, more congenerum.

Folia radicalia primigena oblonga, in petiolum longum decurrentia, indivisa, basi tamen quasi lyrata ob petiolum passim dentibus ceu rudimentis pinnularum instructum, inæqualiter, grosse et obtuse serrata, utrinque valde villosa. Folia radicalia secundaria sæpe incisa, petiolorum appendiculis crebrioribus et majoribus, ita quidem ut jam composita videantur lobo terminali maximo. Eadem omnino ratione, dum dentes petiolares augentur numero et magnitudine, folium vero decrescit et profundius secatur, caulina fiunt verè pinnata, pinnis inciso-serratis, adscendendo angustioribus, magis elongatis, et minus dentatis: summa autem denuo evadunt simpliciora et ramea nonnulla communiter lineari-subulata et simplicia extant. Villositas quoque adscendendo sensimque minuitur, at nunquam penitus evanescit et ad exortum foliorum imprimis adparet.

Pedunculi elongati, subtilissime striati, flores versus pubescentes, præsertim sub ipso calyce.

Calyx communis constat foliolis usque ad 12 lanceolatis, acutis, integris, capitulo brevioribusque, post florescentiam reflexis.

Flosculi quinquefidi radiantes; tubo longo pubescente, laciniis obtusis, extimâ majore, duabus interioribus minoribus, omnibus constanter saturate cæruleis.

Stamina corollulas, styli stamina superant.

Receptaculum paleaceum, paleis subulatis, membranaceis, brevibus vix longitudine germinum.

Semina membrana angustissima, alta lacera coronata, et in medio setis quinque atropurpureis instructa; quæ setæ florente planta corollulas æquant, postea magis adhuc elongatur.

Communis in graminosis sylvaticis per totum hunc tractum. Floret Junio, Julio. ♂

Obs. Autores quamvis multa de Scabiosâ ochroleucâ et columbariâ disseruerint, immo icones adjecerint, differentiam tamen ſufficientem vix ullam hucuſque exhibuiſſe mihi videntur. Foliorum polymorphia et radicalium caducitas compluribus, præferente facem ipſo magiſtro, anſam præbuit. Scabioſis, quæ hic in computum veniunt, ochroleucis folia radicalia pinnato-compoſita, cæruleis autem eadem indiviſa tribuendi. Recentiores vero, animadvertentes ambas haſce notas, ſcilicet cum foliis radicalibus compoſitis flores ochroleucos, et cum foliis radicalibus ſimplicibus flores cæruleos nequaquam ſemper conjunctos eſſe, iverunt in partes, alii e. g. Cl. Jacquin florum colorem, alii e. g. Cl. Roth foliorum formam urgentes. Mihi quidam neutrum conſtituendæ differentiæ ſpecificæ aptum et ſufficiens eſſe videtur; qualis fortaſſe differentia ex curatiore floris et imprimis ſeminum obſervatione et comparatione obtineri poterit. In Tauricâ Cherſoneſo cum floribus conſtanter ochroleucis, ſemina coronula latiore, ſetiſque brevioribus helvolis, et in Caucaſo Caſpico cum floribus ſemper cæruleus, ſemina coronula anguſtiſſima, ſetiſquæ prælongis atro-purpureis obſervavi; inter utriuſque autem folia tam radicalia quàm caulina nullam prorſus differentiam animadverti. An idem etiam alibi, imprimis in Europa occidentali obtinet?

IX. ALCHEMILLA Pubescens. Lamark.

A. Foliis lobatis, lobis cuneiformibus antice crenatis, ſubtus ſericeo-villoſis.

Alchemilla pubeſcens foliis lobatis ſubtus pubeſcentibus, caule villoſo, floribus pedicellatis. Lamark. Tabl. Enc. Bot. n. 1703.

Alchemilla alpina β hybrida. Linn. ed Reiſch. I. p. 349.

Alchymilla minor hirſuta cineritia Italica. Barrel. Ic. 728. bona.

Lecta in montibus altioribus ſiccis circa Kurt-Bulak. Floret Junio. ♃.

Obſ. Præter pubem ſericeam, imprimis foliorum paginæ inferioris et calycum, ab Alchemilla vulgari differt ſtaturâ humiliore et foliis profundius lobatis angulo inciſionis acutiſſimo,

ſimo, adeo ut lobi ſæpius imbricati appareant; qui cæterum obtuſiſſimi ſunt, margine laterali integerrimo et antice crenis paucioribus obtuoſiribuſque quàm in A. vulgari.

X. ONOSMA. Echioides.

O. caule ſupernè ramoſo, foliis lanceolato-linearibus hiſpidis utrinque veridibus, fructibus erectis, corollis calyce duplo longioribus.

Onoſma echioides. Jacq. Auſtr. 3 p. 52. tab. 295.

Borkhauſen apud Römer Nov. Repert. Bot. I. pag. I.

Reliqua Synonyma autorum inter hanc et inſequentem ſpeciem diſpertienda.

β. Onoſma altiſſima: alta, ramoſiſſima, foliis latioribus.

Onoſma orientalis. Pallas Itin. 2 p. 734. tab L?

Congenerum maxima, cujus var. β interdum ad quadripedalem altitudinem excreſcit. facie ſymphyti.

Radix perennis, longa ſimplex, cortice rubro.

Caulis communiter ſeſquipedalis aut bipedalis, ſupernè ramoſiſſimus, pilis ſive ſetulis pellucidis, erectis, creberrimis, è puncto calloſo oriundis quàm maximè hiſpidus, ſicut tota planta.

Folia lanceolata-linearia, aut in var. β lanceolata, ſetulis pellucidis rectis hiſpida, attamen utrinque viridia; inferiora baſin verſus attenuata, ſuperiora ſeſſilia baſi ſubcordata.

Racemi verſus ramorum apices folioſi, ſecundi, juniores convoluti, ætate evolvendi.

Bracteæ foliis ſimiles, niſi quod breviores et minores.

Flores magnitudine Symphyti.

Calyx ad baſin uſque fiſſus, foliolis linearibus, valde hiſpidis.

Corolla generis, calyce duplo longior, ochroleuca.

Antherae baſi connexæ mediantibus hamulis baſeos lateralis, ferè corollæ longitudine.

Stylus ſimplex, corolla longior.

Semina lævia, intra calyces perſiſtentes et erectos.

 Non

Non infrequens in pafcuis et collibus apricis. Var β. altiffima occurrit in arenofis maris Cafpici ad oras rivuli Buam Floret. Junio. ♃

Obf. Onofmæ fpecies Rofficas, quæ mihi hactenus innotuerunt fubnecto, quarum eas, quæ Caucafo Cafpico peregrinæ funt, Afterifko notavi.

* ONOSMA Tinctoria. Nova.

O. caule fupernè ramofo, foliis lanceolata-linearibus hifpidis utrinque viridibus, fructibus erectis, corollis calycem fuperantibus.

Præcedenti nimium affinis differt ftaturâ plerumque humiliore et habitu fimpliciore.

Radix cortice fanguineo inquinante.

Folia præcedente nonnihil anguftiora.

Flores minores, corollis duplam calycis longitudinem non attingentibus, pallidis, defloratis citrinis; fauce paullo magis anguftatâ quàm in antecedente.

Occurrit in Tauriæ pafcuis circa Kertfch et Jenikale, nec non in Ucraniâ et in planitiebus Tanaicenfibus. Floret cum prædecente. ♃.

Obf Forfitan ad hanc fpectat: Cerinthe echioides. Scop. Carn ed 2. n. 197. Contra ad præcedentem potius refero: Onofmam foliis lanceolatis hifpidis, fructibufque pendulis. Gmel. Sib. 4. p. 76. tab. 40; nec non Anchufam luteam majorem C. B. Moris. Hift. 3. p. 442. Sect. 11. tab. 27. fig. 1. Præcipuam à præcedente differentiam fuppeditat florum magnitudo et proportio ad calycem; de quâ differentiâ nihil in autoribus reperio.

* ONOSMA Simplicissima.

O. caule fubfimplici, folii lineari-lanceolatis hispidis fubtus-albido pilofis, fructibus erectis.

Onofma fimpliciffima foliis confertis lanceolato-linearibus pilofis. Linn. ed Reich. 1 p. 396.

Onoſma ſibirica caulii ſimplicibus, foliis lanceolato-linearibus piloſos-hiſpidis, racemo cernuo. Lamark Tabl. Enc. Bot. n. 1836.

Radix perennis, cortice exteriore nigricante.

Caules nonnulli, baſi aliquantum declinati, ſtatim adſcendentes, vix dodrantales, pilis rigidulis albidis parum extantibus et multo exilioribus ac mitioribus quàm antecedentium obſiti.

Folia lineari-lanceolata admodum anguſta, ſupra viridia, pilis caulinis ſimilibus è punćto minimo calloſo originem ducentibus adſperſa, ſubtus vix niſi nervo dorſali pilis hirto, reliquâ autem ſuperficie pube tenui incumbente albidâ et ſplendente veſtitâ.

Racemi pauci ad apicem caulis, forma præcedentiam, braćteis pedunculis calycibuſque magis hirtis quàm caulis et folia.

Corolla formâ, colore, et ferè magnitudine Onoſmæ echioidis.

Habitat in Sibiriâ, regione Tanaicenſi, Ucraniâ. Floret Majo, Iunio ♃.

Obſ. Subſequenti affinis cum præcedentibus confundi nequit, ſtaturâ et pube diverſiſſima.

XI. ONOSMA Taurica.

O. caulibus ſimplicibus è baſi multicipe, foliis lineari-lanceolatis, hiſpidis, utrinque albo-piloſis, frućtibus erećtis.

Onoſma Taurica. Pallas, Tableau de la Tauride p. 47.

Differt ab antecedente caullium baſi multicipe ſubligneſcente; foliis radicallibus numeroſioribus magiſque congeſtis, omnibus utrinque pilis albis crebrioribus et majoribus veſtitis; racemo plerumque tantum unico terminali, corollis paullo majoribus ſaturate luteis.

Frequens in Tauriæ collibis apricis circa Karaſſubaſac et Sympheropolin, paſſim quoque in montoſis Caucaſi Caſpici occurrit. Floret Majo, Iunio. ♃. ♄.

XII. ECHIUM Asperrimum. Lamark.

E. ſtrigoſiſſimum ramoſum, ſpicis lateralibus conjugatis.

Echium aſperrimum caule ramoſo piloſiſſimo, corollis calyce longioribus, ſtaminibus exertis. Lamark Tabl. Enc. Bot. n. 1854.

Echium foliis hirtis lingulatis, calycibus pungentibus, ſpicis laxis paniculatis. Hall. helv. n. 604.

Echium flore albo. epit. 738.

Echium orientale majus et aſperius flore leucophæo Tournef. cor. p. 6.

Commune in incultis ad vias, pagis; antea in Tauriâ quoque copioſè viſum. Floret Junio, Julio. ♃.

Obſ. Icon Camerarii totum plantæ noſtræ habitum exprimit: imprimis racemos conjugatos qui characterem ſpeciei optimum et conſtantiſſimum ſupeditant, bene exprimit; male igitur Linnæus Camerarii plantam ad ſuum Echium italicum retulit, quod a noſtro differre præter alia hoc ſatis probat, quod Echium ſuum italicum mitius et lævius eſſe Echio vulgari Linnæus afferat. (vide Hort. Upſ. p. 35.) Echium altiſſimum. Jacq. Fl. Auſt. noſtro maximè affine eſt, et fortaſſe unum idemque.

XIII. LYSIMACHIA Orientalis. Lamark.

L. racemis terminalibus, petalis obtuſis erectis, foliis petiolatis, imis obovatis, ſuperioribus lanceolatis.

Lyſimachia orientalis racemis terminalibus, ſtaminibus corolla brevioribus, foliis lanceolatis ſubpetiolatis. Lamark Tabl. Enc. Bot. n. 1972. Dict. E. B. III. p. 570.

Lyſimachia ſpicata purpurea minor. Bux. C. 1. p. 22. tab. 33. pygmæa.

Radix perennis, fibroſa, brevis, dilutè purpuraſcens.

Caulis erectus, ſimplex, pedalis, angulatus, infernè purpuraſcens, lævis.

Folia flaccida, glabra, glauceſcentia; ima obovata in petiolum attenuata, ſuperiora lanceolata acutiora brevius petiolata margine undulata.

Racemus

Racemus terminalis, interdum quoque nonnulli laterales, ſimpliciſſimi, elongati, compacti, formâ Veronicæ ſpicatæ.

Pedicelli florentis plantæ breviſſimi, feminiferi elongati ſemunciales.

Bracteæ ſubulatæ, minutæ, formâ foliolorum calycinorum.

Flores vix Veronicæ ſpicatæ floribus majores.

Calyx 5-partitus, ſegmentis lineari-lanceolatis è viridi in ſpadiceum vergentibus.

Corolla monopetala, roſea, calyce longior; tubo breviſſimo; limbo profundè 5-fido, laciniis erectis conniventibus, obtuſis.

Filamenta alba. Antheræ flavæ corollam ſupereminentes.

Stylus in florente ſtaminibus brevior, in fructu perſiſtit et elongatur.

Stigma ſimplex minimum.

Capſula exactè ſphærica, magnitudine Piſi, purpuraſcens, lævis pedicellata, ſtylo ſetaceo coronata.

Creſcit hinc inde in planitie Kubenſi, circa torrentem Samur, ad vias et nemorum margines. Floret Majo, Iunio. ♃.

XIV. CAMPANULA Caucasica. Nova.

C. (capſulis obtectis, calycis ſinubus reflexis) capſulis trilocularibus cernuis, cauliculis erectis paucifloris, foliis obovatis crenatis ſcabris, radice repentè.

Planta perennis, pygmæa.

Radix tenuis, albida, tranſverſaliter ſub terrâ prorepens, et per intervalla folia cauliculoſque promens.

Caules erectiuſculi, ſimplices, tenelli, plerumque digito minores, interdum vix ſeſquiunciam longi, pubeſcenti-ſcabri, ſicut tota plantula.

Folia radicalia obovata, in petiolum attenuata, crenata, hirta, Violæ tricoloris foliis analoga; caulina alterna, minora, anguſtiora et acutiora.

Flos terminalis ſæpe ſolitarius; tum quoque aliqui laterales, pedicellati, primum erecti, mox nutantes, non magni.

Calyx 1. -fidus, ſegmentis ſubæqualibus, acutis, margine ciliatis, alternis, reflexis germen obtegentibus.

Corolla

Corolla cærulea, tubo cylindrico calyce duplo longiore, limbo angusto, laciniis triangularibus.

Squamulæ antheriferæ dilatatæ, breves loco filamentorum, ut solent in hoc genere.

Antheræ oblongæ.

Stylus columnaris, corolla brevior. Stigma trilobum.

Capsula cernua, trilocularis. Semina subrotunda, nitida, rufescentia.

In abruptis sterilissimis lapidosis circa Kurt-Bulak lecta. Floret Junio, Julio. ♃.

Obs. Videtur Campanulæ saxatili proxima, nec a Camp. dichotoma aliena; suæ tamen speciei. Dignoscenda radice late repentè, cauliculos exiguos simplicissimos promente; pube exili; floribus parvis; capsulis trilocularibus cernuis. In floribus quinta pars numeri sæpe deficit; calyce scilicet 8-fido, cor 4-fida, staminibus quatuor.

XV. SALSOLA Kali.

S. herbacea patula hirta, foliis subulatis apice spinosis; calycibus solitariis bracteatis, fructiferis explanatis coloratis.

Salsola Kali herbacea decumbens, foliis subulatis spinosis, calycibus marginatis axillaribusque. Linn. ed Reich. I. p. 624.

Salsola Kali autorem. Pallas it. 1. p. 489. n. 105.

Communis in salsis et subsalsis. Floret Julio, Angusto. ⊙

* SALSOLA Tragus.

S. herbacea glabra, ramis erectiusculis, foliis subulatis apice spinosis, calycibus solitariis bracteatis, fructiferis explanatis hyalinis.

In maritimis Tauriæ circa Kertsch et Jenikale lecta. Floret cum præcedente. ⊙

Obs. Utraque species S. Kali et Tragus occurrit in Tauriâ, ubi solicitè observandi et comparandi mihi occasio fuit. Affinitas quidem utriusque summa, neque vero tanta, ut ad unam eandemque speciem referri queant. Duæ quondam ab anti-

antiquioribus Botanicis ſpecies huc pertinentes deſcriptæ fuerunt; altera foliis longioribus, altera vero magis communis foliis brevioribus et craſſioribus. Hanc III. Linnæus ad S. Kali, illam ad S. Tragum retulit. Equidem, quantum colligere licet ex deſcriptionibus imperfectis et rudibus iconibus, contrarium opinor; ſcilicet illam propriè ad S. Kali Linn. et hanc ad ejuſdem S. Tragum pertinere. Quod in S. Kali ſtylus ſit trifidus, et in Trago bifidus nunquam obſervavi; verum in utrâque conſtanter vidi bifidum. Uberiorem utriuſque deſcriptionem Flora noſtra Taurica exhibebit.

Salſolæ Sodæ in Tauriâ antehac obſervatæ definitionem emendatam ſubnecto:

* SALSOLA Soda.

S. herbacea glabra, ramis adſcendentibus, foliis elongatis ſemiteretibus, calycibus ſubbinatis bracteatis, fructiferorum foliolis dorſo tranſverſaliter carinatis.

In maritimis ſubhumidis Tauriæ circa. Kertoch et Jenikale lecta. Floret Auguſto ⊙

XVI. SALSOLA Rosacea.

S. herbacea patula, foliis linearibus teretibus inermibus, calycibus ſubſolitariis bracteatis, fructiferis explanatis coloratis.

Salſola roſacea herbacea, foliis ſublatis mucronatis calycibus explanatis. Linn. ed. Reich. I. p. 624?

Kali humile alis purpureis florem roſaceum mentientibus, Buxb. Cent. 1. p. 9. tab. 14. fig. 2.

Salſola foliis ternis, floribus ſubſtratis, flores æquantibus. Gmel. Sib. 3. p. 96. n. 75.

Radix annua, ſimplex, albida.

Caulis erectus mox ſpithamæus, mox pedalis et ultra; ramis inde a baſi alternis, patentibus, validis.

Folia linearia, breviora et craſſiora quam in S. Kali, baſi præſertim floraia latiora, omnia inermia glabra.

Flores axillares ſitu et formâ S. Kali.

Calyces quoque fructiferi excrescentes in membranam latam coloratam spadiceam aut fuscam.

Semen ut in S. Kali et Trago.

Frequens ubique in squalidis salsugineis imprimis planitiei ad Cyrum fluvium. Floret Augusto, Septembri. Semina vero autumno perficit. ⊙

Obs. Folia, ne acuminata quidem, in Systemate Linnæano nullo jure dicuntur mucronata; quod fortasse Cl. Pallas in Itin. Vol. 1. p. 489. ansam præbuit quærendi: anne varietas S. Kali pumila, quam describit, sit Salsola rosacea. Linn.?

XVII. SALSOLA Glauca. Nova.

S. fruticosa erecta glabberima, foliis semiteretibus filiformibus, calycibus solitariis bracteatis, fructiferis explanatis hyalinis.

Kali orientale fructicosum altissimum, florum staminibus purpureis. Tournef. Corr. p. 18.

Kali fruticosum spicatum. Buxb. Cent. 1. p. 8. tab. 13?

Frutex ramosus inter sesquipedalem et quadripedalem altitudinem varians.

Caulus erectus, lævissimus, albidus; ramis elongatis, adscendentibus, fragilibus.

Folia filiformia, semiteretia, inermia, glaberrima cum nebulâ glaucâ sicut tota planta; inferiora sesquiuncialia, floralia breviora et sicut in S. Kali, Soda et aliis, basi complanata flosculosque amplectentia.

Inflorescentia, florumque forma et magnitudo ut in antecedentibus; floribus scilicet versus ramorum ramulorumque extremitates alternis, axillaribus, solitariis, inter bracteas duas folio florali substantiâ et formâ similes.

Calyx pentaphyllus erectus; foliolis ovatis, acutis, concavis, dorso viridibus, margine albo-membranaceis.

Filamenta eleganter purpurea, calycem vix æquentia.

Antheræ oblongæ, acutæ, basi bifidæ, filamentis concolores.

Stylus bifidus, staminibus brevior. Stigmata simplicia.

Calyces fructiferi dorso in membranam explanatam, margine vix laceram albidam excrescunt.

Semen

Semen ſitu et formâ S. Kali.

Lecta hinc inde in abruptis ſteriliſſimis circa Kurt-Bulak. Floret Julio. ♄.

Obſ. Synonymon Buxbaumii ſupra allegatum ad noſtram potius pertinet, quàm ad Salſolam fruteſcentem Pallaſii, ad quam ver Cl. trahit Itin. 1. p. 488. Nota Salſolam arboreſcentem Linn. ignoro, credo tamen â noſtrâ S. glaucâ differe.

XVIII. SALSOLA VERMICULATA.

S. fructicoſa ramoſiſſima, foliis imis ſetaceis brevibus, floralibus minutiſſimis; calycibus ſolitariis bracteatis ſparſis, fructiferis explanatis hyalinis.

Salſola vermiculata: fruteſcens, foliis ovatis acutis carnoſis: Linn. ed Reich. 1 p. 627.

Salſola virmiculata fruticoſa, floribus ſpicatis aternis ſolitariis Loefling it p. 129 * Pallas it. 1. p. 488. n. 103.

Kali fructicoſum Ericæ folio Buxb. C. 1. p. 8. tab. 14. fig. 1.

Salſola foliis ternis floribus ſubſtratis, teretiuſculis ovatis, flore brevioribus. Gmel. Sib. 3. p. 98. n. 76.

Fruticulus pedalis aut ſeſquipedalis denſe ramoſus, cortice cinereo fiſſo.

Folia ad imos ramos ſparſa; ſetacea, brevia, inermia, caduca; floralia manutiſſima, concava, forma et ſubſtantia ſquamularum calycinarum, niſi quod minora paulloet auctiora.

Floſculi minuti, ad cujuſvis folii axillam ſolitarii, ſparſi, numeroſiſſimi et ramos totos obtegentes, plerique abortientes.

Bracteæ duæ minimæ, orbiculatæ, obtuſiſſimæ, concavæ, virides, margine tenuiſſimè membranaceo, floſculum ſuum obvolventes.

Calyx clauſis; foliolis oblongis, acutiuſculis, membranaceis, albicantibus cum apiculo tenuiſſimè purpuraſcente.

Glandulæ nectariferæ in fundo calycis, ſtaminibus interjectæ.

Stamina quinque; filamentis brevibus; antheris oblongis, acutis, baſi bifidis, flavis, calycem excedentibus.

Germen ſubrotundum; ſtylus rectus ferè calycie longitudine; ſtigmata ſimplicia.

Calyces fructiferi dorſo in membranam explanatam ſublaceram hyalinam excreſcentes.

Semen ſpirale, in gyros nonnullos convolutum.

Frequens in ſalſugineis aridis, præſertim planitiei ad Cyrum et inde verſus portum Baku uſque ad ſaxum Beſchbarmak: Gratiſſimum Camelis pabulum. Floret Auguſto; ſemina Novembri perficit. ♄

Obſ. Folia inferiora adeo ſunt caduca, ut mihi, ineunte Julio planta primum viſa, planè aphylla apparuerit. Poſtea in herbariis vidi ſpecimina circa Aſtrachanum alibique ad Wolgam, iisdem in locis collecta, ubi è Perſiâ ſeptentrionali redux adultam plantam copioſiſſimam ipſe videram; quæ ſpecimina ad imos ramos manifeſtè folioſa extabant. Hæc foliorum genuinorum raritas et caducitas, jam Buxbaumio cauſa fuit dicendi floſculos creberrimos foliorum conſpectum eripere. Flores juniores cum ramulis ſæpe pube levi cinereſcunt, quæ in adultioribus evaneſcit. Vidi quoque in herbario Cl. et Amic. Stephani, Profeſſoris Moſquenſis, cujus de re herbariâ Ruthenâ merita ſupra in contextu hujus Opuſculi commendavi, ſpecimina circa Saratow lecta, quæ quantum é ſiccis colligere licebat, a noſtris radice annuâ et ſtaturâ pygmaeâ, nec ulla prorſus alia nota differebant. Kali bacciferum granuloſum. Gmel Sib. 3. pag. 99. n. 77. Var. II. tab. 21. fig. 2. nihil aliud erit quam Salſola vermiculata noſtra, qualis emarcida et morboſa ſæpe conſpicitur inter Aſtrachanum et Kiſlar, unde Gmelinus ſua ſpecimina acceperat.

XIX. SALSOLA HYSSOPIFOLIA.

S. herbacea patula, foliis linearibus planis ſubpubeſcentibus, calycibus glomeratis ebracteatis, fructiferorum foliolis ſpina dorſali uncinata. Salſola hyſſopifolia. Pall. it. 1. p. 491. n. 107. tab. H. fig. 1.

Planta annua, Salſ. muricatæ affinis.

Radix tenuis, fibroſa, albida.

Caulis erectus, circiter pedalis, purpuraſcens ſicut Salſ. proſtratæ, inde à baſi ramoſiſſimus; ramis patulis, elongatis, attamen minus quam in Sals. muricata.

Folia

Folia formâ Salf. muricatæ, fed paullo breviora minufque pubefcentia, margine non ciliata. Lana vaga ad foliorum floralium axillas.

Flofculi axillares, plerumque plus duobus, glomerati, magnitudine et formâ Salf. muricatæ.

Styli 2 filiformes, longi, purpurei.

Calycum fructiferorum foliola fornicata germinique fupernè incumbentia, dorfo exerunt fpinulam innocuam, albidam, uncinatam.

Semina corticata, formâ Salf. muricatæ.

Occurrit in fubfalfis ad arundineta fluvii Cyri. **Legi** Novembri florentem. ⊙

Obf. A. Salfola muricata facillimè dignofcitur calycum fructiferorum fpinula uncinata.

Salfolam hyffopifoliam à noftrâ certè diverfam defcribit Nocca apud Ufteri Annal. Botan. 6. p. 60.

XX. SALSOLA MURICATA.

S. herbacea, erecta pilofa; ramis filiformibus; foliis linearibus planis; calycibus binis ebracteatis, fructiferorum foliolis fpina dorfali recta.

Salfola muricata: fruticofa patula, ramulis hirfutis, calycibus fpinofis. Linn. ed Reich. 1. p. 628.

Salfola fedoides. Pallas it. 1. p. 492. App. n. 108. tab. 1. fig. 1. 2. et III. p. 630. tab. M. fig. 3.

Chenopodium maritimum, ramulis virgatis. Buxb. C. 3. p. 27. tab. 49.

Occurrit rarius in fterilibus fubfalfis. Floret autumno. ⊙

Obf. Plantæ in Tauriâ frequentiffimæ defcriptionem Flora Taurica noftra continebit. Hic definitio fpeciei emendata fufficiet.

XXI. SALSOLA PROSTRATA.

S. fublignefcens, ramis adfcendentibus, foliis linearibus planis pilofis, calycibus glomeratis ebracteatis, fructiferis explanatis coloratis.

Salfola

Salſola proſtrata fruteſcens, foliis linearibus piloſis inermibus. Linn. ed Reich. 1. p. 627. Pall. it. 1. p. 490. n. 106. tab. G. e.

Salſola proſtrata. Loefl. it. p. 131. * optima deſcriptio.

Kali foliis Linariæ tomentoſum. Buxb. C. 1. p. 10. tab. 16. peſſima icon.

Communis in collibus apricis inter Kubam et Schamachiam, ſolo vix ſalſo. Floret Julio, Auguſto. ♄. ♃.

Obſ. Gallis ſæpe inficitur, unde Buxb. loc. cit. aſſerit: dari plantas quæ florum loco gerant globulos axillares lanuginoſos, interquos tubuli corniculati; cui deſcriptioni gallarum forma exactè reſpondet.

XXII. SALSOLA FRUTICOSA.

S. fruticoſa ramoſiſſima glabra; foliis oblongis obtuſis carnoſis; calycibus ternis ebracteatis petiolaribus, fructiferorum foliolis convexis dorſo æqualibus.

Salſola fruticoſa: erecta, fruticoſa, ſoliis filiformibus obtuſiuſculis. Linn. ed Reich. 1. p. 627.

Frequens cum S. vermiculata occurrit, quam ſtaturâ et ramificatione æmulatur; cæterum Salſolæ altiſſimæ, imprimis quoad infloreſcentiam affinis. Floret cum Salſ. vermiculata. ♄

Obſ. Salſola baccifera ramoſiſſima. Gmel. Sib. 3. pag. 99. n. 77. Var. I. tab. 21. fig. 2. videtur eſſe Salſola fruticoſa, qualis adparet autumnum verſus aut vernali tempore triſtines vegetans, aut reviviſcens. Saltim Gmelini ſpecimina ex eo ipſo tracta allata ſunt, ubi equidem Salſolam fruticoſam copioſiſſimam vidi è Perſia ſeptentrionali redux, Februario a. 1797.

XXIII. SALSOLA ALTISSIMA.

S. herbacea glabra, ramis erectis, foliis filiformibus carnoſis, calycibus ternis ebracteatis petiolaribus, fructiferorum foliolis convexis dorſo æqualibus.

Salſola altiſſima: herbacea erecta ramoſiſſima, foliis filiformibus acutiuſculis baſi peduncuIiferis. Linn. ed Reich. 1. p. 625.

Paſſim

Paſſim in ruderatis, ad foſſas et vias obvia. Floret cum præcedente. ⊙

Obſ. Plantæ ſatis cognitæ definitionem tantum hic inſerendam putavi.

XXIV. SALSOLA MARITIMA.

S. herbacea patula glabra, foliis linearibus obtuſiuſculis carnoſis, calycibus glomeratis ebracteatis axillaribus, fructiferorum foliolis convexis dorſo æqualibus.

Chenopodium maritimum: foliis ſubulatis ſemicylindricis. Linn. ed Reich. 1. p. 612.

Occurrit in glareoſis inundatis ſalſugineis. Floret Julio. ⊙

Obſ. Stirpem hanc a Sals. altiſſimâ genere nullo jure ſeparari, alii jam monſtraverunt. Chenopodium, Salſola, Anabaſis, nec non Polycnemum, ſaltim quoad ſpecies nuper autoritate Cl. Pallaſii ad hoc genus relatas, genera ſunt adeo affinia, ut limites naturales difficilimè conſtituantur. Salſolæ bracteis propriis carentes, imprimis eae, quarum calyx fructifer dorſo in membranam haud excreſcit, ob ſemen corticatum lenticulare, Chenopodio propriores ſunt. Polycnema Pallaſii meliori jure ad Anabaſin referuntur, quam ad Polycnemum Linn. flore et fructu ab illis alienum. In ſpeciebus hactenus a Syſtematicis ad Anabaſin relatis foliola calycina variant et numero et formâ; eſſentia vero generis potiſſimum quærenda eſt in ſemine ſolitario cochleato, intra membranam liquore quodam oleoſo turgidam verticaliter poſito; quæ nota etiam in polycnemis Pallaſii obſervatur.

Quibus perpenſis ſequentes caracteres genericos naturæ conſentaneos eſſe arbitror:

a. CHENOPODIUM: Cal. 5-phyllus fructiferi foliolis conniventibus dorſo æqualibus. Cor. o. Semen unicum corticatum.

Huc referænde: Salſola fruticoſa, altiſſima, ſativa.

b. SALSOLA: Cal. 5-phyllus, fructiferi foliolis medio tranſverſaliter carinatis appendiculatis. Cor. o. Semen unicum tectum, ſitu horizontali. (ſæpe ſpirale.)

 Salſolæ

Salfolæ in duas phalanges difcedunt, quarum altera complectitur Sodas: flofculis bracteatis, femine cochleato; altera Chenopodoideas; flofculis ebracteatis, femine corticato. Ad hanc pertinent: Sals. proftrata, muricata, hyffopifolia et aliæ; ad illas vero: Sals. Kali, Tragus, rofacea, Soda, glauca, vermiculata.

c. Anabasis: Cal. 2-5-phyllus. Cor. o. (nifi calycem dicas.) Semen unicum fpirale, intra membranam liquore oleofo turgidam verticale. Huc pertinent: Polycnemum triandrum, oppofitifolium, fclerofpermum. Pall. Salfola Echinus. La Billard.

d. Polycnemum folo Polyc. arvenfi Linn. abfolvitur.

XXV. ANABASIS APHYLLA.

A. pentandra, ebracteata; calyce pentaphyllo minimo, foliolis tribus dorfo appendiculatis, caule articulato aphyllo.

Anabafis aphylla, articulis emarginatis. Linn. ed Reich. 1. p. 628.

Salfola baccifera Salicorniæ facie. Gmel. Sib. 3. p. 101. n. 78.

Kali bacciferum Salicorniæ facie. Buxb. C. 1. p. 11. tab. 18.

Pallas it. 1. p. 493. App. n. 109. β. quoad flores.

Occurrit haud raro in diftrictu falfo ad Cyrum ufque ad oppidum Baku. Floret Julio, Augufto. Novembri femina fert. ♄

Obf. Frutex ficut foliis ita bracteis quoque deftituitur. Autores huic fpeciei tribuunt calycem triphyllum et corollam pentapetalam; fed me quidem judice nullo omnino jure. Quod illis calyx vocatur, nihil aliud eft, quam membrana ex veri folioli calycini dorfo originem trahens, ætate aucta et dilatata; ficut accidit in plerifque Salfolis, quibus nemo hucufque calycem duplicem vel corollam adfcribere voluit. Quod in Anabafi aphylla, è foliolis calycinis duo tali membrana deftituantur; et quod ipfa membrana in reliquis foliolis erectior fit, atque refpectu folioli fui major; id fanè haud fufficit, ſufficit, ad conſtituendum caracterem, congeneribus et Salſolæ ſpeciebus affinibus planè alienum. Florem optimè deſcripſit Cl. Pallas loco ſupra citato.

XXVI. ANABASIS Echinus.

A. pentandra, calyce pentaphyllo, foliolis omnibus dorſo appendiculatis, foliis ſubulatis, ramis ſpineſcentibus.

Salſola Echinus: fruticoſa glabra, foliis ſubulatis muticis, ſpinis divaricatis. La Billardière Syr. dec. 3.

Anabaſis ſpinoſiſſima Linn. ſecundum Vahl. Symb.

Suffrutex humilis, ſpinoſus.

Radix perennis, lignoſa, valida.

Caules baſi ligneſcentes, divaricati.

Rami adſcendentes, herbacei, circiter pedales, virgati, ſtriato-angulati, juniores pubeſcenti-ſcabri, ætate glabri et glauceſcentes ſicut tota planta. Ramuli rigidi, apice ſpineſcentes, ad angulum rectum patentes, breves et adſcendendo decreſcentes.

Folia ima filiformia, ſemiteretia, mucrone minimo inermi terminata, foliorum ſecundariorum breviorum faſciculo axillari.

Folia ramea breviora, baſique latiora; floralia bracteis ſimilia iiſdemque paullo longiora.

Flores utrinque ad exortum ramulorum oppoſiti, in ipſis autem ramulis alterni, axillares, formâ florum Polycnemi triandri Pall.

Bracteæ duæ ovato-lanceolatæ, acutæ, carinatæ, margine albo-membranaceæ, calycem includentes.

Calyx pentaphyllus, bracteis paullo brevior; foliolis ovatis, concavis, albo-membranaceis, genitalia obvolventibus.

Stamina quinque, vix calyce longiora.

Filamenta apicem verſus purpuraſcentia.

Antheræ purpureæ, oblongæ, baſi bifidæ, filamentis infra apicem affixæ, polline flavo.

Germen ſubrotundum.

Styli duo, recti, conniventes. Stigmata ſimplicia.

 Calycis

Calyces fructiferi foliola excrefcunt in membranam explanatam, margine apicis vix laceram, e viridi dilute purpurafcentem, in duobus foliolis intermediis paullo anguftiorem more Salfolæ.

Semen unicum cochleatum, intra membranam liquore oleofo turgidam verticale.

In collibus aridiffimis, abruptis, fubfalfis paffim obvia. Lecta copiofius Circa Kurt-Bulak; in collibus quoque a Cyro fluvio non diffitis hinc inde vifa. Floret Julio. Semina Octobri perficit. ♄.

Obf. Icon la Billardierii loc. cit. noftram plantam bene exhibet; verum defcriptio peccat eo, quod ramos apice fpinefcentes fpinas nuncupat.

XXVII. ANABASIS TRIANDRA.

A. triandra, calyce triphyllo, foliolis dorfo æqualibus, foliis lineari-fubulatis recurvis.

Polycnemum triandrum. Pallas it. 1. p. 483. App. n. 95. tab. D. fig. 2. et rab. E. fig. 1.

In glareofis falfugineis. Floret Julio, Augufto. ⊙

Obf. Hanc in Tauriâ quoque obviam, cum ftirpibus Tauricis ulterius defcribam. Male eam cum Polycnemo arvenfi, toto cœlo diverfo, conjungit. Lamark. Tabl. Enc. Bot. n. 439.

XXVIII. ANABASIS OPPOSITIFOLIA.

A. pentendra, calyce diphyllo, foliolis dorfo æqualibus, foliis linearibus carnofis obtufis.

Polycnemum oppofitifolium: caulibus erectis, foliis femicylindricis tormentofo-glaucis, inferioribus oppofitis, floribus pentandris. Lamark. Tabl. Enc. Bot. n. 441.

Polycnemum oppofitifolium. Pall. it. 1. p. 484. tab. E. fig. 2.

Chenopodium maritimum Sedi foliis teretibus. Buxb. C. 1. p. 21. tab. 31. fig. 1.

Occurrit

Occurrit cum præcedente, at ſolum magis ſalſum amat. Floret Julio. ☉

Obſ. Planta in Florâ noſtrâ Tauricâ amplius deſcribenda.

XXIX. ERYNGIUM CAERULEUM. Nov.

E. foliis radicalibus cordato-ovatis, caulinis palmatis; involucri foliolis quinis ſubulatis baſi utrinque ſpinulâ minimâ armatis, paleis mucronatis.

Eryngium cæruleum ſtellatum montis Libani. Munting. Aardgew. Cap. 127. p. 455. n. 5.

Eryngium foliis radicalibus ovatis crenatis petiolatis, capitulis pedunculatis. Gronov. orient. n. 76.

Eryngium Syriacum ramoſum, capitulis minoribus cæruleis. Moris. Hiſt. 3 Sect. 7. p. 166. tab. 37. fig. 13.

Radix perennis, ſimplex, foris nigra, intus albida.

Caulis dodrantalis, raro pedalis, tenuis, baſi ſimpliciſſimus, ſupernè in ramos paucos ſubdichotomos patulos diviſus, teres, leviſſimè ſtriatus, colore pulchre amethyſtino pictus quo tota hæc planta elegans ob imo ad ſummum ornatur.

Folia radicalia petiolata, aut indiviſa, aut obſoletè lobata, cordato-ovata, crenata, obtuſiuſcula, caduca; caulina compoſita, ſubpalmata; ſegmentis lanceolatis dentatis, dentibus ſubulatis apice pungentibus et ſæpe latere quoque ſpinula una alterave armatis.

Capitula pedunculata, parva, ſubgloboſa, ad ramorum apices.

Involucrum pentaphyllum, capitulis quadruplo longius; foliolis ſubulatis, apice et utrinque ſupra baſin ſpinula munitis, cæterum integerrimis.

Paleæ ſubulatæ, mucronatæ, integræ.

Abundat in collibus apricis ſiccis provinciæ Schirvanienſis, quibus colore ſuo amoeniſſimum ſæpe aſpectum conciliat. Floret Junio, Julio. ♃

Obſ. Affinis Eryngio tricuſpidato Linn., at paleæ certè et conſtanter integræ. Elegans planta et præſtanti colore inſignis. Synonyma Gronovii et Moriſonii Linnæus ad Eryng.

 tricuſ-

tricuſpidatum trahit; revera autem noſtro magis congrua ſunt, imprimis Moriſonii Icon, quæ manifeſtè paleas integras et in ſimplicem mucronem deſinentes, exhibet.

XXX. BUPLEURUM EXALTATUM. Nov.

B. involucro univerſali partialique ſubpentaphyllo minimo; foliis omnibus linearibus, inferioribus elongatis; caule erecto paniculato.

In pratis montanis et collibus ſiccioribus non infrequens. Floret Junio, Julio. ♃

Obſ. Stirps Tauriæ indigena, cum Tauricis deſcribenda, a compluribus affinium ſtaturâ altiore et radice perenni diſtinguenda, Bupl. rigido affinis: diſcrepat foliis etiam imis anguſtiſſimis ceu gramineis, involucris et involucellis exilioribus.

XXXI. CACHRYS MICROCARPOS. Nov.

C. foliis tripartito-decompoſitis, foliolis ſetaceis, ſeminibus glabris ſtriis crenatis.

Cachris ſemine fungoſa ſulcato aſpero minore foliis Peucedani. Moris. hiſt. 3. p. 267. Sect. 9. tab. 1. fig. 1.

Cachris foliis Peucedani, ſemine fungoſo, ſulcato, aſpero, minori. Tournef. Corr. p. 23.

Planta umbellifera perennis ſpecioſa.

Radix deſcendens, fibroſa.

Caules è radice ſæpe nonnulli, ſeſquipedales aut bipedales, ramoſi, ſtriati, glabri et laetè virrides ſicut tota planta.

Rami patentes, elongati, ſubdiviſi.

Folia ſupradecompoſita; petiolis trichotomis, ita ut trichotomiæ ramus medius ſemper debilior exiſtat. Petioli primæ ſecundæ et in primordialibus foliis etiam tertiæ diviſionis teretes ſunt atque ſtriati, ſubſtantiâ ramorum; ulteriores autem ſunt ſubſtantiâ foliolorum.

Foliola ad petiolos ultimæ diviſionis plerumque terna, interdum bina, rarius ſolitaria, ſubulata, circiter unciam longa, rigida, convoluta, mucrone apicis minimo.

Folia

Folia caulina pro more umbellatarum adſcendendo evadunt ſimpliciora; floralia demum ſimpliciter ternata, immo ſuprema ſimplicia, forma foliolorum involucri, niſi quod latiora.

Umbellæ ad caulis ramorumque apices, nec non aliquæ laterales, oppoſitæ, pedunculatæ; ſæpe quoque pedunculi tres quatuorve umbellam ſuſtinentes ex eodum puncto originem ducunt.

Pedunculi ſulcati, rigidi.

Radii umbellæ à 3 ad 8, ſtriati, flavicantes.

Involucrum univerſale 3-8-phyllum, pro numero radiorum umbellæ; foliolis ſubulatis, breviſſimis.

Radii umbellulæ circiter 8.

Involucrum partiale univerſali conſimile, ſed brevius.

Floſculi omnes fertiles.

Petala acutiuſcula, involuta, lutea, colore fœniculi.

Stamina concoloria, patentiſſima.

Styli in florente minimi, in fructu reflexi.

Stigmata minima, ſubgloboſa.

Fructus ſubgloboſi, minime marginati, glabri, ſtriis crenatis, exarati.

Semina vix corticata: magnitudine circiter ſeminum Conii maculati, valde convexa, ſingula lineis elevatis crenatis quinque inſtructa; ſaporis aromatici acris, linguam inſtar piperis rodentia.

Creſcit hinc inde ad margines lapidoſos rivulorum in montanis ſylvaticis inter Kubam et Schamachiam. Floret Julio. Semina Septembri perficit. ♃.

Obſ. planta facie Peucedani, fructu Conii, attamen Cachrydi Siculæ proxima, quo cum Linnæo recentiores etiam ſynonymon Moriſonii retulerunt. Differt manifeſtè ſeminibus minoribus glabris.

XXXII. HERACLEUM PYRENAICUM. Lamark.

H. foliis ſimplicibus lobatis pubeſcentibus, ſubtus incanis, floribus uniformibus, ſeminibus margine ciliato-ſcabris.

Heracleum pyrenaicum: foliis ſimplicibus cordato-palmatis ſuprà glabris, ſubtùs tomentoſis albis, ſeminibus orbicularibus. Lamark. Dict. Enc. Bot. 1. p. 403.

Radix generis.

Caulis erectus, vix pedalis, craſſus, ſulcato-angulatus, leviter pubeſcens.

Folia fici foliis analoga, ſimplicia, ſubpalmata, cordato-ovata, lobata; inferiora petiolata, petiolo teretiuſculo hirto, lobis ſupernis oblongis, medio trifido, lateralibus baſi productis, margine omni denticulis minutiſſimis argutis.

Folia caulina ſenſim minora, ſuprema lobis integris, brevius petiolata, petiolis in vaginam amplam membranaceam ſtriatam definentibus. Cæterum folia omnia utrinque molliter pubeſcunt, ſubtùs autem tomento tenuiſſimo albicant.

Umbellæ ad caulis ramorumque apices formæ ſolitæ in hoc genere, pedunculis et radiis ſulcatis hirtis.

Floſculi albi, uniformes.

Semina formæ communis, ſcabra, diſco lineis tribus elevatis viridibus et quatuor ſulcis ſpadiceis notato, margine ſubtiliſſimè ciliato.

Odor totius plantæ fortis, non ingratus, ad odorem Rutæ accedens.

Lectum paſſim in lapidoſis ad rivulos ſylvaticos, et in abruptis altioribus inter Kubam et Schamachiam, imprimis circa Kurt-Bulak. Floret Majo; Junio ſemina perficit. ♂

Obſ. de ſynonymo Cl. Lamarkii nullus dubito, quamvis folia ſuprà glabra dicantur. Cæterum ad Heracleum alpinum proximè accedit, tamen ſtaturâ, pube et denticulis marginalibus crebris minutiſque, nec non aliis notis, ſatis diſtinctum.

XXXIII. SESELI CUNEIFOLIUM. Nov.

S. foliolis cuneiformibus furcatis ſuperioribus oblongis integris, ſeminibus villoſis.

An Crithmum ſive fœniculum majus, odore Apii. Moris. hiſt. 3. Sect. 9. p. 290. tab. 7. ic.?

Radix perennis, valida, ramoſa.

Caulis erectus, ſæpe bipedalis, teretiuſculus, glaberrimus

cum

cum nebulâ glaucâ ſicut tota planta; ſupernè ramoſus, ramis elongatis, rigidis, nudiuſculis.

Folia Galbani foliis ſimilia, pinnato-compoſita; petiolis elongatis, forma et ſubſtantia ramulorum.

Foliola ad petiolos ultimæ diviſionis ternata aut pinnato-quinata, cuneiformia, margine laterali integerrimo quaſi abſciſſo; antice inciſo-ſerrata ſæpe trifurca, ſubcarnoſa et admodum rigida.

Folia ſuperiora ſimpliciora; foliolis anguſtioribus acutioribus; ramea obſoleta, ſcilicet ad vaginæ anguſtæ apicem rudimentis tantum foliolorum obſoletorum.

Umbellæ pedunculatæ, ferè formâ Seſeleos annui. Radii umbellæ 6 aut 8 craſſi.

Involucrum univerſale nullum.

Umbellulæ glomeratæ, hemiſphæricæ, parvæ.

Involucrum partiale breviſſimum, connate-monophyllum, multidentatum.

Floſculi ſubſeſſiles.

Petala inflexa, acuta, non emarginata, ex albido virideſcentia.

Antheræ albæ. Styli florentis breviſſimi.

Semina ovata-oblonga, ſulcata, villis albis hirta.

Odor totius plantæ fortis Apii hortenſis.

Floret Julio paſſim in abruptis lapidoſis altioribus circa Kurt-Bulak et verſus fontes rivulorum Pirſagat et Sugaite. ♃

Obſ. Habitus totus imprimis umbellæ de genere nullum prorſus dubium relinquit. Foliolis latuiſculis cuneiformibus furcatis facillimè a congeneribus diſtinguitur. Synonymon Moriſonii dubium eſt.

XXXIV. PASTINACA PIMPINELLIFOLIA. Nov.

P. foliis pinnatis, foliolis inciſo-ſerratis, inferioribus ſubrotundis, ſuperioribus oblongis.

Paſtinaca orientalis foliis eleganter inciſis. Tournef. Corr. p. 23. Buxb. C. 3. p. 17. tab. 27.

Radix perennis.

Caulis erectus, raro pede altior, tenuis, ſulcatus, pubeſcens, ſupernè in ramos aliquot floriferos diviſus.

Folia

Folia pinnata, petiolata, utrinque pubefcentia, villis exiguis incumbentibus, Pimpinellæ faxifragæ foliis fimilia et vix majora.

Foliola in inferioribus foliis fubrotunda, incifo-ferrata, bafi antica quafi excifa, poftica in lobum producta. In fuperioribus foliis foliola fenfim evadunt longiora et partius incifa, in fupremis autem linearia integra extimo majore, eadem planè ratione quà in Pimpinella faxifraga.

Radii umbellæ conftanter fex.

Involucrum univerfale fæpe foliolo uno alterove obfoletò caduco; partiale fæpe tryphyllum, foliolis minutis fubulatis, unilateralibus.

Corollulæ inflexo-emarginatæ faturate luteæ.

Fructus generis, compreffo-planus, non marginatus.

Lecta in graminofis fertilioribus altioribus circa Kurt-Bulak. Floret Julio. ♃.

Obf. ftaturâ humili, et foliis Pimpinellæ analogis, a congeneribus illico diftinguenda.

XXXV. PIMPINELLA PEREGRINA.

P. foliis pinnatis pubefcentibus, inferiorum foliolis cordato-ovatis ferratis, fuperiorum oblongis; radiis umbellæ numerofis feminibufque hirtis.

Pimpinella peregrina: foliis radicalibus pinnatis crenatis, fummis cuneiformibus incifis, umbellis nubilibus nutantibus. Linn. ed Reich. 1. p. 724.

Pimpinella peregrina. Jacq.

Radix fufiformis, fimplex, alba foris nigricans, biennis.

Caulis erectus, bipedalis et nonnunquam altior, bafi fimplex, fupernè divifus in ramos elongatos floriferos, teres, fubtiliter ftriatus, pubefcens bafi infignius.

Folia pinnata, petiolata, utrinque leviter pubefcentia ficut tota planta.

Foliola ima cordato-ovata, ferrata, at vix incifa five obfoletiffimè tantum triloba; bafi exteriore productiore.

Foliola fuperiora longiora et anguftiora, magis incifa; fuprema linearia; in ramis præfertim floriferis elongata et pauciora.

Umbellæ

Umbellæ ante anthefin terram refpicientes, florentes fructiferæque erectæ funt.

Radii umbellæ numerofi, interiores breviores, hirti, imprimis umbellulam verfus.

Involucrum univerfale rarius foliolo caduco inftruitur, fæpiffimè deficit.

Flofculi numerofi, pedicellati, albi, formæ communis.

Semina formâ et magnitudine pimpinellæ faxifragæ, pilis albis infigniter hirta.

In graminofis fylvaticis non infrequens. Floret Junio. ♂.

Obf. affinis Pimpinellæ magnæ; nec umbellæ ante florefcentiam nutantes caracterem diftinctivum efficiunt, qui potius ab hirfutie herbæ et imprimis feminum, porrò a foliolis rameis elongatis, denique ab umbellâ ampliffimâ radiis numerofis videtur effe repetendus.

XXXVI. PIMPINELLA Danaa.

P. involucro utroque, foliis radicalibus decompofitis incifis, caulinis minimis, feminibus læviffimis.

Danaa aquilegifolia. Allion Pedem. II. p. 34. n. 1392. tab. 63.

Liguſticum alterum Belgarum. Lob ic. 786. Liguſticum alterum herbariorum. Tabern. ed. Bauh. p. 205. Jc.

In fylvarum altiorum umbrofis haud rara. Floret Junio, Julio. ♃

Obf. feminum forma, ad quam in umbelliferis fecundum recentiorum plerorumque opinionem potiffimum refpiciendem eft, ftirpem hanc ad Pimpinellam referre, fuadet; involucrum tamen utrumque et femina læviffima ambiguam reddunt. An igitur ob hanc unicam fpeciem novum genus a Cl. Allionio conftitutum, in fyftemata introducendum? Cæterum Icon Allionii non ex optimis eft, imprimis foliola jufto majora fiftit. In Tauriæ umbrofis altioribus æque frequens olim a me obfervata, cum ftirpibus Tauricis defcribetur.

XXXVII. STA-

XXXVII. STATICE Echinus.

S. caule fuffruticofo foliofo ramofo, foliis confertiffimis acerofis, fcapis breviffimis floribus alternis obtectis.

Statice Echinus: fcapo paniculato, foliis fubulatis mucronatis. β. Limonium græcum Juniperi folio. Tournef. Corr. p. 25. Linn. ed Reich. 1. p. 755.

Tragacantha tota fpinofa et Echium referens unde Echinus herbaceus dicta. Moris. hift. 2. Sect. 2. App. tab. 23. fig. 7.

Radix, ut folet in hoc genere, lignofa, valida, emittens denfo cæfpite caules lignefcentes breviffimos, denfè ramofos et hemifphæram femi-pedalis interdum diametri referentes, quæ foliis confertiffimis echinata nomen triviale fibi obtimo jure vindicat.

Folia antiquiora emarcida ad bafin ramorum retrorfum imbricata; juniora recta, acerofa, femi-teretia, mucronato-pungentia, punctis minimis albicantibus pubeque vix confpicua glaucefcentia.

Scapi vel fi malis pedunculi ex apice laterali ramulorum folitarii, fimplices, digito breviores, fæpe vix fefquiunciales, toti bracteis floribufque alternis imbricatis obtecti. Squamulæ aliquot fubulatæ, albo-membranaceæ ad exortum fcaporum.

Bracteæ ad fingulos flores tres; quarum exterior florem feffilem fcapumque ceu vagina amplectitur, oblonga, acuta, carinata, viridis, margine apiceque anguftè membranaceo; reliquæ duæ totæ albo-membranaceæ, fubæquales, calycis tubum obvolventes, dorfo vena purpurea.

Flores figuræ communis, Limonii floribus duplo majores.

Calyx tubo cylindrico, bafi parum pubefcente, fupernè ftriis quinque purpureis in limbum excurrentibus; limbo explanato, fcariofo, indivifo margineque crenato, crenis decem fubtiliffimè crenulatis, alternis fcilicet iifquæ venis dorfalibus carent minoribus.

Petala calyce longiora, rofea cum vena media obfcuriore, oblonga, apicem verfus latiora, obtufa.

Stamina vix calycem excedentia.

Piftillum

Piſtillum generis, ſtylis 5, ſtigmatibus incraſſatis.

Creſcit hinc inde in abruptis ſteriliſſimis altioribus. Lecta circa fontes rivuli Sugaite. Floret Junio. ♃ ♄

Obſ. definitio Linnæana a noſtrâ plantâ adeo eſt aliena, ut duas diverſas ſpecies ſub St. Echino Linn. latere, credibile ſit; ſcilicet noſtram, quæ tanquam var. β proponitur, et alteram ad quam Linnæus definitionem ſuam adornavit, et quam exprimit: Limonium ceſpitoſum, foliis aculeatis. Buxb. Cent. 2. p. 18. tab. 10. Buxbaumiana enim planta, ſcapo elongata paniculato gaudens a noſtra manifeſtè differt, quippe quæ conſtanter ſcapos ſimpliciſſimos, raro digitalis longitudinis promit, qui ferè toti floſculis alternis obteguntur, vel potius ex eorundem pedicellis breviſſimis conſtant; matureſcentibus enim ſeminibus, immo ſi planta intra chartam exſiccatur, flores cum ſuis pedicellis decidunt, et eo ipſo longe maxima pars ſcapi ex totidem articulis conflata, evaneſcit.

XXXVIII. STATICE lyrata. Nova.

S. caule herbaceo, foliis radicalibus lyratis caule ſubnudo ſimpliciſſimo, ſpicis elongatis glomeratis.

Statice caule nudo ſimpliciſſimo, ſpicis florum ſeſſilibus alternis, foliis radicalibus ex ſinuato pinnatis. Gmel. Sib. 2. p. 224. tab. 91. f. 2.

Radix annua, tenuis, albicans.

Scapus radicalis, digitalis, rariſſimè dodrantalis, aphyllus, baſi teretiuſculus, apicem verſus acutangulus, villioſiuſculus. (Non raro ſcapi aliquot ex eadem radice oriuntur.)

Folia radicalia plura, petiolata, lyrata; lobis ſinubuſque obtuſis; cæterum ſubvilloſa, viridia.

Spica terminalis, et ſæpe nonnullæ laterales ſtrictæ, compoſitæ ex ſpicis partialibus breviſſimis glomeratis, inferioribus alternis, ſuperioribus minoribus atque contiguis.

Bractea ſub unaquaque ſpica partiali, ſive glomerulo minuta, ſubrotunda cum acumine, viridis, margine tenuiſſimè albo-membranaceo.

Floſculi parvi, paleis oblongis, apicem verſus parum dilatatis margineque villoſis interſtincti. Bractea ſingulis floſculis

culis propria, subquadrangula, obtuissima, viridis calycem obvolvens.

Calyx tubo cylindrico striato basi pubescente, limbo dilatato albo scarioso crenulato et aristis quinque subtilissimis ejus substantiæ munito.

Petala obtusa, rosea.

Stamina vix corolla longiora; filamentis albis, antheris lutescentibus.

Non rara in argillosis subsalsis ad mare Caspium. Floret Junio. ⊙

Obs. Affinis St. lobatæ et sinuatæ. Differt ab utrâque floribus minutis, et ab hâc caule non alato. Descriptionem satis curatam cum Icone plantæ totum habitum exprimentè dedit Gmelinus loco cit. nihilominus species hæc a recentioribus Systematicis neglecta est.

XXXIX. ALLIUM SAXATILE. Nov.

A. scapo nudo tereti, foliis semiteretibus spatha bivalvi brevi, staminibus subulatis corolla longioribus.

Cepa scapo nudo tereti inani, foliis semicylindricis, capitulis spissis multifloris. Gmel. Sib. 1. p. 63. tab. 16. fig. 1. 2.

Lectum in abruptis sterilissimis circa Kurt-Bulak. Floret Septembri. ♃

Obs. In Tauriæ collibus saxosis frequens est, ubi petala constanter lactea sunt cum nervo dorsali viridi: in Caucaso Caspico autem nervus medius semper dilutè purpurascit. An Allium Sibiricum Linn.? sed stamine dicuntur corolla breviora, quod non in nostro.

XL. EPILOBIUM ROSMARINIFOLIUM. Curtis Bot. Mag.

E. staminibus declinatis, foliis sparsis linearibus denticulatis caulibusque subpubescentibus, racemis foliatis.

Lysimachia siliquosa speciosa angustifolia. J. Bauh. hist. II. lib. 21. p. 907.

Caules e radice perenni multi, in cespitem collecti, adscendentes, bipedales et ultra, purpurascentes, pubescentes; pube

pube ſubtiliſſima et nonniſi facie nebulæ glaucæ, qua caulis inumbratres videtur apparentè.

Rami floriferi nonnulli verſus apicem caulis.

Folia ſparſa, linearia, anguſtiora quàm E. anguſtifolii, ſeſſilia, margine denticulis minimis ceu glandulis purpuraſcentibus; cæterum viridia, cum nebulâ fugaci glauca præſertim verſus baſin. Folia juniora ſæpe faſciulatim ex axillis primariorum inferiorum naſcuntur.

Racemi terminales ſimplices, foliati; ſcilicet bracteis è quarum axillis flores prodeunt foliis ſimilibus, minoribus tantum.

Pedicelli breviſſimi.

Calyx tetraphyllus, germini lineari infero albo-tomentoſo oblique inſiſtens; foliolis linearibus, acutis, purpuraſcentibus, pubeſcentibus, petalis paullo brevioribus.

Petala calycinis foliolis interjecta, patentiſſime, æqualia; unguibus brevibus, limbo roſeo, ovato, obtuſo, ſed minus quàm in Ep. anguſtifolio.

Stamina 8 declinata, ferè corollæ longitudine; filamentis tenuibus cæruleſcentibus; antheris luteus, polline cæruleo.

Stylus ferè longitudine ſtaminum, amethyſtinus, baſi villis albis pubeſcens; ſtigma minutum trilobum.

Frequens ad margines lapidoſos rivulorum e. g. rivi Pirſagat propè Veteram Schamachiam. Floret Auguſto. ♃

Obſ. ab Epilobio anguſtifolio Linn. diſcedit foliorum anguſtiâ et denticulis lateralibus minutiſſimis crebrioribus, pube, bracteis ſive foliis floralibus, petalorum limbo paullo minus dilatato, ſtigmate minuto. Propius videtur accedere ad Epilobium Dodonæi. Allion. Pedem. 1. n. 10:6. et ad Epil. anguſtifolium. Lamark Dict. 2. p. 374.

XLI. POLYGONUM SALSUGINEUM. Nov.

P. floribus ſubpentandris trygynis axillaribus, foliis lineari-ſubulatis, ſeminibus calycibus duplo longioribus.

An Polygonum maximum longiſſimis cauliculis et foliis. Moris. hiſt. 2. Sect. 5. p. 591. Tournef. Inſt. p 510.?

Radix

Radix annua, ſimplex, alba.

Caulis pedalis et altior, erectus, inde à baſi ramoſiſſimus; ramis patentiſſimis, elongatis, gracilibus, teretibus, ſubtiliſſimè ſtriatis, glabris ſicut tota planta; geniculis purpuraſcentibus.

Folia diſtantia, lineari-ſubulata, margine revoluto, lævia, caduca, ſtipulis impoſita.

Stipulæ amplexicaules, breves, margine ciliato laceræ, florales folio deſtitutæ.

Floſculi axillares, ſubterni, breviſſimè pedunculati, polygoni avicularis floribus minores.

Calyx viridis, monophyllus, quinquedentatus, dentibus ſubæqualibus, minutis, anguſtis, rectis.

Stamina mihi quinque viſa, longitudine calyce, antheris globoſo-didymis, flavis.

Germen oblongum, ſtamina æquans. Stylus nullus. Stigmata tria, minima, albida.

Semen nigricans, triquetro-pyramedale, calyce duplo longius.

Creſcit paſſim in argilloſis ſchiſtoſos, bitumine et ſale imbutis, ſteriliſſimis. Copioſius in vicinia veteris Schamachiæ. Floret Auguſto, Septembri. ☉

Obſ. abundè differt a Polygono aviculari ejuſdemque varietatibus ſive ſpeciebus affinibus anguſtifoliis. Semine pyramidali Stelleris affine. Staminum numerus difficillimus eſt obſervatu, ob flores exiguos et antheros admodum fugaces.

* RUTA LINIFOLIA Linn.

R. foliis lanceolatis ſubſeſſilibus glabris, paniculæ ramis pubeſcentibus, petalis ovatis.

Ruta linifolia: foliis lanceolatis indiviſis. Linn. ed Reich. II. p. 266.

Ruta ſylveſtris linifolia hiſpanica. Barrel. Icon. 1186. Tournef. Inſt. p. 257.

Caules è radice perenni nonnulli, ſimplices, circiter pedales, infernè glabri, ſupernè pubeſcentes.

Folia

Folia lanceolata, integerrima, glabra, glaucefcentia, opaca fcilicet punctis pellucidis nullis notata qualia in affinibus obfervantur, omnia fubfeffilia.

Paniculæ fubcorymbofæ rami infignius pubefcunt.

Calycina foliola fubrotunda, lævia, parva, viridia.

Petala ovata, obtufa, lutea, dorfo fubtus præfertim apicem verfus viridia.

Stamina decem, bafi interiore parum villofa.

Lecta in Tauriæ collibus ficcis. Floret Majo, Junio. ♃

Obf. plantam hanc et duas affines Rutam dauricam et perforatam, in Caucafo Cafpico non repertas, ideo hic tracto, ut Rutæ foliis fimplicibus gaudentes, quot quot mihi hucufque innotuerunt, ordine fuo definitæ melius innotefcant, earumque difcrimina facilius agnofcantur. Rutæ perforatæ fpecimina è Sibiriâ accepta, mecum benevolentiffimè communicavit Cl. et amiciffimus Stephan.

* RUTA DAURICA.

R. foliis lanceolatis fubfeffilibus glabris, paniculæ ramis lævibus, petalis ovatis. Ruta foliis fimplicibus alternis. Gmel. Sib. IV. p. 176. n. 95. tab. 68. fig. 1.

Harmala montano Daurica perennis, multicaulis, Polygalæ folio, flore albo et luteo. Amman. Ruth. n. 91. 92.

Peganum dauricum: foliis indivifis. Linn. ed Reich. II. p. 421.

Planta præcedenti affinis ftatura et habitu.

Caules fimplices toti glabri.

Folia antecedente paullo minora et auctiora, punctis pellucidis notata, margine obfoletè ferrulata, tamen lævia.

Paniculæ rami glabri, graciliores quàm præcedentis.

Flores pauci, antecedentis magnitudine.

Calycina foliola minuta.

Petala ovata, obtufa, lutea, extus dorfo fæpe viridia.

Stamina certè decem, filamentis parum villofis.

Capfula formâ rutæ, at fæpe tantum triloba.

Variet. β. floribus albis Ammannus defcribit. loc. cit.

XLII. RUTA villosa. Nov.

R. foliis lanceolatis petiolatis cauleque villoſis, petalis oblongis, filamentis baſi lanatis.

Ruta orientalis Linnariæ folio, flore parvo. Tournef. Corr. p. 19. Buxb. C. 2. p. 30. tab. 28.

Habitus antecedentium.

Folia brevibus petiolis inſiſtentia, integerrima, utrinque pubeſcentia ſicut caulis, punctis pellucidis notata, inferiora breviora petiolis longioribus.

Panicula ſubcorymboſo, floribus minoribus numeroſis, confertis.

Calycina foliola minutiſſima, ut vix nudo oculo conſpiciantur.

Petala anguſta, parva, tota lutea.

Stamina decem, baſi interiore villis albis hirſutiſſima et quaſi lanata.

Habitat in provinciæ Schirvan editis apricis circa ſaxum Beſchbarmak, ubi olim Buxbaumius quoque collegit. Floret Junio, Julio. ♃

Obſ. huic affinis videtur: Ruta fruticuloſa, foliis integris ſeſſilibus ovato-lanceolatis ramiſque piloſis. La Billardiere Syr. Dec. 1.

* RUTA perforata. Nov.

R. foliis ovato-lanceolatis petiolatis cauleque glabris petalis oblongis.

Præcedentibus altior et ramoſior, tota glaberrima et ſubglauceſcens.

Folia ovato-lanceolata, latiora quam antecedentium, petiolis brevibus inſiſtentia, pellucido-punctata.

Flores formâ et magnitudine Rutæ villoſæ, ſed filamenta baſi villis multo parcioribus veſtita nec lanata.

Habitat circa mare Baikal, unde primus attulit B. Sievers, Pharmacop. Kiacktenſis. ♃

XLIII. DIAN-

XLIII. DIANTHUS plumarius?

D. floribus ſubſolitariis, ſquamis calicinis ſenis ovato-lanceolatis calyce triplo brevioribus, corollis multifidis.

An D. plumarius: floribus ſolitariis, ſquamis calicinis ſubovatis breviſſimis, corollis multifidis fauce pubeſcentibus. Linn. ed Reich. II. p. 336?

Radix perennis lignoſa ceſpitans; à quâ foliorum faſciculi, et

Caules nonnulli, circiter ſemipedales, erecti, ſimplices, ſæpe uniflori, quandoque ramo uno alterove florifero laterali, teretiuſculi, glabri.

Folia ut ſolent in hoc genere linearia, acuta, glabra margine ſcabriuſcula; caulina ſuperiora brevia ſubmembranacea ſtriata, quorum ſupremum par interdum ſub ipſo calyce ſitum ſquamas calycinas mentitur.

Flores magnitudine D. Caryophylli.

Calyx cylindricus longus; dentibus acuminatiſſimis quaſi ariſtatis, ſquamis baſeos ſex, tubo triplo brevioribus, ex ovato lanceolatis cum acumine medio parum elongato; exteriores duæ minores.

Petala incarnata, immaculata; unguibus calyce paullo longioribus, limbo ovato antice ciliato-lacero.

Stamina et Piſtillum formæ communis.

Color totius plantæ è viridi in glaucum vergens.

Creſcit hinc inde in editis inter Kubam et Schamachiam. Floret Julio. ♃

XLIV. CUCUBALUS Royeni. Nov.

C. racemo verticillato, pedunculus oppoſitis breviſſimis multifloris, petalis bifidis, foliis ovato-lanceolatis ſubundulatis.

Cucubalus foliis amplexicaulibus, floribus verticillatis pedunculatis erectis. Royen Liegdbat. p. 448.

Radix biennis.

Caulis erectus, ſimpliciſſimus, ſeſquipedalis aut bipedalis, teretiuſculus, villoſo-viſcidus, imprimis ſupernè.

Folia ovato-lanceolata, ſubundata, pleraque integerrima, nonnulla interdum dente elongato laterali; caulina ſemiamplexicaulia, adſcendendo breviora; floralia brevia, acuminata; omnia utrinque villoſo-viſcida ſicut tota planta.

Racemus terminalis ſimplex longus verticillatus, ſcilicet pedunculo utrinque ex axillâ folii floralis breviſſimo, è quo

Flores quinque aut plures, ſinguli pedicello recto vix calycis longitudine inſiſtentes, magnitudine circiter Silenes nutantis.

Calyces cylindrici, decemſtriati, quam maxime viſcidi.

Petala albida, ſemibifida coronula deſtituta.

Genitalia corolla longiora.

Calyces fructiferi ventricoſiores et ſubcavati.

Semina minima, punctulato-ſcabra, atra.

Lectus paſſim in planitie Kumükorum. Floret Majo. ♂

Obſ. affinis Cucubalo viſcoſo, ad quem Linnæus, quamvis dubius, Synonymon Royeni retulit. Differt foliis latioribus floribuſque erectis nec undique decumbentibus.

XLV. SILENE SUFFRUTESCENS. Nov.

S. (floribus ex dichotomia caulis) caulibus baſi ligneſcentibus, foliis ſpatulato-lanceolatis, calycibus cylindricis 10-ſtriatis; petalorum limbo bipartito, margine baſeos utrinque unidentato.

Suffrutex humilis, divaricatus, habitu Silenes fruticoſæ.

Caulis baſi ligneſcens, ramis floriferis herbaceis elongatis teretibus, ſupernè paniculatis ſubdichotomis, leviter pubeſcentibus ſicut tota planta.

Folia oblonga, apicem verſus latiora, in medio acutiuſcula, margine ſcabra; caulina inferiora ſæpe foliorum ſecundariorum faſciculo axillari; ramea ſuperiora longiora et anguſtiora.

Pedunculi longi, cylindrici, ſtriis decem obſcurioribus inſtructi, pubeſcentes.

Petala bipartita, è purpureo in ſpadiceum vergentia, denticulo utrinque minuto marginali ſupra coronulam faucis.

Genitalia

Genitalia formæ communis in hoc genere, calyce longiora.

Antheræ luteo-virides.

Capſula ovata, intra calycem pedunculata; quo fit, ut calyx fructus clavatus videatur.

Occurrit hinc inde ad latera collium lapidoſorum inter Kubam et veterem Schamachiam. Floret Junio, Julio. ♄ ♃

Obſ. Planta affinis Cucubalo fruticuloſo Pall. at coronuta faucis etiam genere differe probat.

XLVI. COTYLEDON SEMPERVIVUM Nov.

C. faſciculis globoſis è foliis cuneiformibus integris margine ciliato-ſcabris; ſcapis radicalibus, panicula oblonga laxa.

Radix perennis, paſſim emittens foliorum roſulas globoſas more Simpervivi.

Folia numeroſa, denſe imbricata, cuneiformia, lata et brevia, angulis lateralibus rotundatis paullo productioribus, margine toto ciliis breviſſimis ſcabro; cæterum carnoſa, vix pubeſcentia.

Scapi radicales ſimpliciſſimi, ſpithamæi, tenues, ſemiteretes, pubeſcentes.

Panicula terminalis, oblonga; pedunculis alternis brevibus, bi-aut trifloris, villoſo-viſcidis.

Bracteæ minutæ, lanceolatæ, obtuiſiuſculæ.

Flores Sidi Telephii floribus paullo majores.

Calyx 5. partitus, villoſo-viſcidus; ſegmentis lanceolatis, longitudine tubi corollæ.

Corolla monopetala, 5. fida, colore Sedi Telephii; tubo germinibus adpreſſo; limbo erectuiſcula, laciniis oblongis acutis.

Stamina 10; filamentis tubo corollæ inſertis eademque brevioribus, antheris parvis globoſo-didymis purpureis

Piſtilla 5 ſupera, oblonga, ſuperne anguſtata; ſtigmate minimo.

Capſulæ 5 polyſpermæ.

Occurrit rarius in ſaxis montium altiorum. Floret Junio. ♃

Obſ. Folia Sempervivi; flores Cotyledonis.

XLVII. STACHYS FRUTICULOSA. Nov.

S. caule lignefcente ramofiffimo divaricato, foliis elliptico-lanceolatis fubintegris, verticillis paucifloris.

Suffrutex humilis, ramofiffimus, divaricatus.

Rami floriferi adfcendentes, vix dodrantales teretiufculi, villis incumbentibus pubefcentes.

Folia omnia uniformia, feffilia, elliptico-lanceolata, obtufa nec medio mucronata, plerumque integerrima et raro tantum obiterque crenata, non magna, eodum modo quo rami pubefcentia.

Flores formâ et magnitudine Stach. rectæ, ad folia fuperiora verticillati, breviter pedunculati; verticillis paucifloris, fæpiffimè bifloris, fcilicet utrinque flore axillari folitario.

Calyx 10. ftriatus, 5-dentatus; dentibus æqualibus mucronatis, mucrone reflexo-patulo.

Corolla alba. Stamina generis.

Semina pleraque abortientia, adeo ut communiter tantum unicum ad maturitatem perveniat, majufculum, læve, rufefcens cum nebulâ obfcuriore.

Lecta in abruptis lapidofis aridiffimis excelfis cum Statice Echino. Floret Majo, Junio. ♄

XLVIII. SCROPHULARIA VARIEGATA. Nov.

S. caulibus bafi lignefcentibus, foliis bipinnatifidis pubefcentibus, racemis elongatis, pedicellis brevibus villis glochidibus hirtis.

Radix perennis, lignofa, multiceps.

Caulis bafi lignefcens divaricatus; cefpitatim emittens ramos numerofos, herbaceos, elongatos, graciles, ftriato-angulatos, pubefcenti-fcabros ficut tota planta.

Folia compofita, bipinnatafida; fegmentis bafeos minoribus, oppofitis, groffe ferratis; terminali maximo incifo-ferrata; fuperiorum lobis lateralibus cum terminali confluentibus.

Racemi terminales elongati, foliofi; pedunculis ad axillas

foliorum floralium alternorum et minutorum folitarius breviffimis; inferioribus trifloris biflorifve, fuperioribus unifloris, omnibus villis breviffimis apice glochidibus hirtis.

Flores paullo minores quàm Scroph. caninæ.

Calycina foliola fubrotunda, obtufiffima, viridia, margine tenuiffimè albo-membranaceo.

Corolla purpurafcens, maculis albis notata-Labium fuperius laciniis duabus brevibus obtufis incumbentibus, altera macula alba. Labium inferius trifidum album, laciniis minimis obtufis.

Stamina generis. Stylus deflexus, labio inferiori incumbens.

Capfula fubglobofo, magnitudine Pifi, ftylo perfiftente et rigefcente acuminata.

Semina minutiffima, punctulato-fcabra, atra.

Crefcit ad margines lapidofos rivulorum, nec non in collibus fterilibus hinc inde. Floret Junio, Julio. ♄ ♃

Obf. Videtur Scroph. caninæ proxima, à quà differt radice perenni caulibus ramofis bafi lignefcentibus, pube caulis et pedunculorum, floribus albo-maculatis.—Speciei affinis Tauricæ definitionem fubnecto.

* SCHROPHULARIA chrysanthemifolia. Nov.

S. foliis bipinnatafidis glabris, fegmentis linearibus incifis, racemis paniculatis, pedunculis fubdichotomis.

Schrophularia orientalis Chryfanthemi folio, flore minimo variegato. Tournef. Corr. p. 9.

Schrophularii pinnata: foliis pinnatis, foliolis incifis, racemis fimpliciffimis terminalibus. Mill. Dict. n. 15?

Haud rara in lapidofis Tauriæ montofæ. Floret Majo, Junio. ♃

Obf. Schrophulariæ lucidæ proxima, et foliis glaberrimis fubcarnofis gaudens.

XLIX. OROBANCHE Phelypæa.

O. caule fimplici unifloro, calyce nudo quinquefido, foliolis tribus fuperioribus erectis conniventibus.

Lathræa phelypæa. Linn.

Phelypæa orientalis flore coccineo. Tournef. Corr.

Radix pro more generis in aliarum plantarum radicibus parasitica.

Caulem promens mox unicum, mox nonnullos. simplicissimos, dodrantales, spadiceos, squamosos, Orobanches majoris caulibus simillimos.

Flos terminales, solitarius, sessilis, speciosus, ringens et ob hanc rationem unaversum inclinatus, Orobanches majoris flore triplo major.

Calyx 5 fidus cauli concolor: segmentis tribus superioribus erectis, conniventibus, oblongis, canaliculatis; duobus inferioribus longioribus et acutioribus antrorsum flexis corollæque tubo sub labio inferiore incumbentibus.

Corolla speciosa, pro more generis extus pube subviscida tenuissima instructa, lutescens; intus holoserico et colore præstantissime coccineo insignis.

Labium superius bipartitum, laciniis obtusis integris; labium inferius planum, tripartitum; laciniis subæqualibus, obtusiusculis. Faux maculis duabus magnis nigerrimis.

Stamina corolla breviora, exteriora interioribus longiora. Antheræ basi bifidæ, bisetæ.

Stylus longitudine staminum; stigma capitato-hemisphæricum.

Capsula generis.

Primum lecta in collibus graminosis circa Fortalitium Alexandrinum et Moscoviense in districtu Caucaso Circassiorum finitimo; postea hinc inde in umbrosis altioribus Dagestani et Schirvani visa. Me quidem judice ad Orobanchen potius quàm ad Lathræam referenda. Floret Majo, Junio. ♃

L. ALYSSUM ALPESTRE.

A. caulibus suffruticosis diffusis, foliis brevibus obtusis incanis, racemis subsolitariis, siliculis ellypticis.

Alyssum alpestre: caulibus suffruticosis diffusis, foliis subrotundis incanis, calycibus coloratis. Linn. ed Reich. 3. p. 233.

Alyssum

Alyssum caulibus fruticulosis diffusis, foliis subrotundis incanis. Gerard. Provinc. p. 352. tab. 13. fig. 2.

Alyssum alpestre. Lamark Dict. Enc. Bot. 1. p. 97. n. 4.

Caules basi sublignescentes, declinati, ramosissimi.

Folia inferiora conferta, basin versus attenuata apice latiora, obtusiuscula, incana; superiora longiora et angustiora.

Racemi terminales elongati simplices.

Flores alyssi montani floribus paullo minores.

Calycina foliola oblonga, concava, forulosa, coloratiuscula, erecta.

Petala lutea, obtusiuscula et vix emarginata.

Stamina 2 breviora basi inter germen squama setacea ipsis dimidio breviore; 4 longiora supra medium utrinque denticulo minimo instructa.

Silicula ellyptica, formâ Drabæ vernæ, stylo persistente coronata, bilocularis; loculis monospermis.

Lectum in abruptis aridis ad Kurt Bulak. Floret Majo. ♃ ♄

Obs. Silicula elliptica characterem distinctivum hujus speciei longe optimum præbet. Specici affinis Tauricæ definitionem subjungo.

* ALYSSUM SUBALPINUM. Nov.

A. caulibus suffruticosis diffusis, foliis subrotundis incanis, racemis subcorymbosis, siliculis orbicularibus.

Non rarum in rupestribus et lapidosis Tauriæ montosæ. Floret Majo. ♃ ♄

Obs. Habitus totus et folia antecedentis, at racemis pluribus ad ramorum apices corymbos efformantibus et siliculorum forma facile dignoscendum: Alysso quoque montano haud mediocriter affine cum eo ab autoribus confusum fuisse videtur; sed racemi etiam Alyss. montani solitarii simplices, calyces non colorati, nec caules basi lignescentes; staminum denique structura diversa. In Alysso subalpino squamulæ 6 subæquales oblongæ apice subtilissimè denticulatæ circa germen connivent; quibus stamina 4 longiora margine laterali, et 2 breviora dorso adnata sunt: in Alysso montano stamina

4 lon-

4 longiora infra medium denticulo laterali inſtruuntur; breviora vero originem ducunt è baſi dorſali ſquamulæ oblongæ apice ſubtiliſſimè denticulatæ, denticulo medio acutiore.

LI. CHEIRANTHUS CUSPIDATUS. Nov.

C. foliis lanceolatis dentatis, caule recto ſimplici, ſiliquis ſtrictis late ancipitibus ſtylo perſiſtente duplo longioribus.

Turritis montana ſiliquis latis. Buxb. Cent. 2. p. 23. tab. 33. f. 1.

In collibus graminoſis ad Sylvarum margines obvius, in Cherſoneſo Tauricâ frequentior. Floret Majo, Junio ♂

Obſ. Cheir. montano. Pall. proximus, foliis dentatis et ſiliquæ ſtylo breviore dignoſcendus. Cheir. montani ſequens erit definitio.

* CHEIRANTHUS MONTANUS.

C. foliis lanceolatis integerrimis, caule recto ſimplici, ſiliquis ſtrictis late-ancipitibus longitudine ſtyli perſiſtentis.

Cheiranthus montanus. Pall. it. 1. p. 496. App. n. 115.

Cheiranthus cornutus: foliis linearibus integerrimis canaliculatis recurvis, floribus ſubſeſſilibus, ſiliquis brevibus longo ſtylo terminatis. Lamark. Dict. Enc. Bot. 2. p. 717. n. 8.

Lectus in Tauriâ ad littora maris arenoſa prope Theodoſiam. Floret Junio. ♂

LII. CHEIRANTHUS ODORATISSIMUS.

C. foliis lyrato-ſinuatis tomentoſis; ſiliquis longiſſimis ancipitibus, ſtigmate bilobo terminatis, caule ſuffruticoſo.

Cheiranthus odoratiſſimus. Pallas Tabl. de la Tauride. App.

Heſperis anguſtifolia V. foliis ſinuato-dentatis ſubruncinatis. Lamark Dict. Enc. Bot. 3. p. 322. n. 5.

Lectus in abruptis ad rivum Pirſagat prope veterem Schamachiam. In Tauriæ cretaceis frequentior. Floret Aprili, Majo. ♄

LIII. LAVA

LIII. LAVATERA BIENNIS. Nov.

L. herbacea, foliis inferioribus cordatis crenatis ſubrotundo-lobatis, ſummis oblongis trilobis lobo medio maximo, pedunculis elongatis ſolitariis, petalis truncatis.

Radix ſimplex biennis videbatur.

Caulis erectus, ſimplex, ſeſquipedalis aut bipedalis, pubeſcenti-ſcaber.

Folia alterna petiolata; inferiora cordato-rotundata, ſubquinqueloba; lobis acutiuſculis ſerratis, baſeos minoribus; ſuperiora magis oblonga; ſuprema lanceolata, triloba, lobis lateralibus minimis: omnia eodem modo quo caulis totaque planta pubeſcenti-ſcabra, nullo tomenti vel canitiei veſtigio.

Stipulæ minutæ, ſubulatæ, integræ.

Pedunculi ex axillas foliorum ſuperiorum ſolitarii, elongati ſcilicet calyce vel quadruplo longiores, tenues, uniflori.

Flores formâ et magnitudine Lavateræ trimeſtris.

Calyx generis; exterior trifidus, lobis ex ovato-orbiculatis, cum acumine medio brevi mutico; interior quinquefidus, ſegmentis ex ovato-lanceolatis.

Petala leviſſimè emarginata et quaſi truncata, formæ communis et colore roſeo Lavat. trimeſtris.

Semina nuda, nullo orbiculi veſtigio formæ communis.

Occurrit paſſim in paſcuis graminoſis et ad agrorum verſuras. Floret Junio, Julio. ♂

Obſ. A. Lavatera trimeſtri defectu orbiculi ſemina tegentis abunde differt: a Lavatera thuringiaca diſcrepat: habitu omnino graciliore, caule è radice veriſimiliter bienni ſæpiſſimè tantum unico ſimpliciſſimo, floribus magis diſtantibus et pedunculis imprimis fructiferis gracilioribus et longioribus, petalis magis conniventibus et vix emarginatis nec profundè exciſis ut in Lav. thuringiaca, denique pube rariori et aſperiori nec molliuſcula vel in tomenti ſpeciem abeunte.

LIV. LATHYRUS ROTUNDIFOLIUS. Nov.

L. pedunculis multifloris, cirrhis dyphillis foliolis ſubrotundis, internodiis nudis.

Occurrit

Occurrit in nemoribus ſub fruticibus quibus innititur. Floret Majo. ♃

Obſ. Elegantem hanc ſpeciem in Tauriæ quoque nemoribus occurrentem curatius in Flora hujus Cherſoneſi deſcribam. Floret tempore vernali cum Orobis. Orobus ſylvaticus foliis circa caulem auriculatis. Buxb. C. 3. p. 22. tab. 41. huic non abſimilis eſt, ſed ſtipulæ in noſtro non ſunt adeo magnæ.

LV. ERVUM TENUISSIMUM. Nov.

E. pedunculis filiformibus ſubunifloris, calicibus breviſſimis.

Haud infrequens in graminoſis apricis. Floret Majo. ⊙

Obſ. Ervo tetraſpermo affine et forte ab autoribus cum eo confuſum: etenim in Tauriâ et per totum tractum Tanaicenſem non rarum, olim in Germaniâ quoque meridionali mihi, ni fallor, occurrit. Ervum tetraſpermum Leers Fl. Herborn. et Lamark Dict. Enc. potius ad noſtram ſpeciem quàm ad E. tetraſpermum Linn. pertinere videtur. Differt ab hoc: foliolis paucioribus, cirrhis longis ſimpliciſſimis, pedunculis apice vix in mucronem protractis, floribus omnium minimis, calycinis dentibus breviſſimis.

LVI. ASTRAGALUS ONOBRYCHIOIDES. Nov.

A. (cauleſcens erectus) pubeſcens, foliis petiolatis, foliolis ellypticis, ſpicis ſubcapitatis longiſſime pedunculatis, calicinis dentibus bracteiſque acuminatis, vexillo elongato, leguminibus oblongis erectis.

Aſtragalus orientalis caneſcens, capitulis Trifolii bituminoſi, flore dilute purpureo. Tournef. Corr. p. 28?

Radix lignoſa, valida.

Caules è ſtipite baſeos ſublignescente aliqui, recti, vix pedales, ſimplices, baſi folioſi pube adpreſſa ſericea caneſcentes, ſupernè nudiuſculi ſulcati virides attamen pubeſcentes.

Folia

Folia alterna, petiolata, eadem pube quâ bafis caulis canefcentia.

Petioli tenues, longi fcilicet foliolorum conjugationibus longo demum a caule intervallo incipientibus.

Foliola a 10-ad 15-juga cum impari, oppofita aut alterna, Aftragali Onobrychis foliolis breviora, ellyptica, parva.

Stipulæ oppofitifoliæ, amplexicaules, albo-membranaceæ, bifidæ; laciniis fubulatis, villofis margineque quafi ciliatis.

Spicæ florum breves, fubcapitatæ, terminales.

Bracteæ fubulatæ, membranaceæ, villofæ imprimis margine, longitudine calycis.

Flores ferè magnitudine et formâ Aftragali Onobrychis.

Calyx albido-villofus intermixtis villis parcis nigris; dentibus acuminatis hirfutis, longitudine tubi.

Corolla purpurea, colore dilutiore quàm Aftr. Onobrychis.

Vexillum alis carinâque duplo longius obtufum. Alæ integræ.

Legumina non racemofa, fed potius in capitulum congefta, erecta, brevia, fubtus gibba, acuminata, villis mixtis albis nigrifque hirta.

Semina reniformia, atra, nitida.

In collibus lapidofis non infrequens, imprimis circa Kurt-Bulak. Floret Majo, Junio. ♃

Obf. Aftragalum Onobrychin refert, quamvis certè diverfus et fpica fubcapitata breviffima, bracteis fubulatis, calycinis dentibus acuminatis, nec non foliis longè petiolatis diftinguendus.

LVII. ASTRAGALUS PILOSUS β FLORE PALLIDO.

A. (caulefcens erectus) villofus, foliolis ellyptico-lanceolatis, fpicis longè pedunculatis brevibus, carina unguiculata, leguminibus oblongis erectis.

Aftragaloides montana incana major albo flore. Barrel. ic. 557.

Radix perennis.

Caules erectiufculi, vix pedales, fimplices, teretes, molliter villofi.

Stipulæ

Stipulæ breves, amplexicaules, acuminatæ.

Foliola circiter 10-juga cum impari, ſæpe alterna, ellyptico-lanceolata, utrinque villoſa, vix incana.

Pedunculi axillares, firmi, elongati, ſulcati; inferiores longiſſimi.

Flores racemoſi, non multi, conferti, brevitur pedunculati.

Bracteæ ſubulatæ, minimæ, villoſæ.

Calyx villoſus pilis mixtis atris raris nec caneſcens; dentibus brevibus ſubæqualibus.

Corolla pallidè floreſcens ad ochroleucum vergens.

Vexillum ovatum, emarginatum, alis paullo longius; alæ ſemiſagittatæ apice obſoletè emarginatæ; Carina baſi laterali utrinque gibba, unguiculo lineari longitudine dentis calycini terminata; quâ notâ hæc ſpecies a compluribus affinium facillimè dignoſcitur.

Legumina racemoſa, imbricata, ferè formâ leguminum antecedentis ſpeciei, oblonga, mucronata, ſupra canaliculata, ſubtus ſubearinata, villoſa.

Semina reniformia, parva, nigro-nitida.

Occurrit cum præcedente et eadem tempora Floret. ♃

* ASTRAGALUS PILOSUS α FLORE SATURATIORE.

A. villoſiſſimis, foliolis ellyptico-lanceolatis, pedunculis folio longioribus, ſpicis cylindricis, carina unguiculata, leguminibus oblongis erectis.

Aſtragalus piloſus: cauleſcens erectus piloſus, floribus ſpicatis, leguminibus ſubulatis piloſis. Linn. ed Reich. 3. p. 526.

Creſcit in collibus ſiccioribus Tauriæ. Floret Majo, Junio. ♃

Obſ. Aſtragalus piloſus α et β, ille in Tauriâ, hic in Caucaſo Caſpico lectus, tot notis conveniunt, ut utrumque ſpecie ſeparare non audeam. Differt α à β villoſitate majore, pedunculis minus elongatis, ſpicis longioribus floribuſque numeroſioribus, bracteis longitudine calycis, dentibus calycinis magis acuminatis, corollis ſaturatius luteis.

LVIII. ASTRA-

LVIII. ASTRAGALUS VARIUS.

A. (caulefcens erectus) pubefcens, foliolis lineari-lanceolatis, racemis elongatis longe pedunculatis, floribus fubfeffilibus diftantibus, leguminibus ovato-oblongis erectis.

Aftragalus varius. S. G. Gmelin Itin. 1. p. 116. tab. 17.

Aftragalus varius: caulefcens erectus, floribus longè fpicatis laxis erectis, ftipulis nigricantibus. Lamark Dict. Enc. Bot. 1. p. 313. n. 21.

Radix perennis, è quâ

Caules nonnulli, erecti, fefquipedales, parum ramofi, teretiufculi, pube incumbente fericea veftiti.

Folia pinnata; foliolis quinquejugis aut ultra, lineari-lanceolatis, acutis, femunciam longis, pubefcentibus ficut caulis.

Stipulæ parvæ, lanceolatæ.

Racemi verfus apicem caulis axillares, elongati caulemque fæpe fuperantes, longè pedunculati.

Flores alterni, minime imbricati, erectiufculi, fubfeffiles, magnitudine nec formâ Aftragali Onobrychis.

Bracteæ minimæ, fubulatæ.

Calyx villofus, villis albis et atro purpurafcentibus mixt ; dentibus 5-fubulatis, fubæqualibus, margine albido-villofiffimis.

Vexillum cæruleum, ex ovato-oblongum, obtufum, alis longius neque vero duplo. Alæ femicordatæ, acutiufculæ, cæruleæ. Carina alis brevior, obtufa, pallida, macula apicis faturàtiore.

Legumina erecta, ex ovato-oblonga, acuta, villofa, non magna, ferè formâ leguminum Aftr. Onobrychis.

Lectus in arenofis ad mare Cafpium circa Tarki et Derbentum. Floret Junio. ♃

Obf. Stipulas caulemque fub illis nigris villis valdè pubefcere et quafi barbatum effe afferit Cl. Lamark loc. cit.: quod fortaffe in culto obtinet; in fpontaneo villi ifti fæpe defunt, nec unquam adeo confpicui, ut definitionem fpeciei intrare mereantur.

LIX. ASTRAGALUS SANGUINOLENTUS. Nov.

A. (ſcapo nudo, abſque caule folioſo) pubeſcens, ſcapis foliis longioribus declinatis paucifloris; foliolis ellypticis ſupra glabris, leguminibus erectiuſculis oblongis arcuatis compreſſis.

Radix perennis, lignoſa, valida, apice multiceps, e qua ceu ex ſtipite ligneſcente foliorum faſciculi ſcapique denſo ceſpite originem ducunt.

Folia humi ſtrata, petiolata; foliolis parvis, circiter 10-jugis cum impari, ex ellyptico ſubrotundis, obtuſiſſimis et leviſſime emarginatis, e viridi glauceſcentibus, ſupra glabris, ſubtus villis incumbentibus pubeſcentibus.

Scapi circiter digitales, foliis longiores, ſupini, teretes, pubeſcentes eodum modo quo foliolorum pagina inferior.

Flores terminales non multi, vix quinis plures, breviſſime pedicellati, erecti, magnitudine et forma florum Aſtr. exſcapi.

Bracteæ ſubulatæ, vix pedicello longiores.

Calyx cylindricus, pubeſcens villis albis nigriſque incumbentibus, 5-dentatus; denticulis æqualibus acutis brevibus.

Vexillum calyce duplo longius, margine apicem verſus reflexo, emarginatum, ex albido dilutiſſime purpuraſcens imprimis baſi ſuperne.

Alæ vexillo haud multo breviores, albæ, apice laterali obſolete emarginatæ. Carina alis brevior, obtuſa, alba, macula apicis magna eleganter cyanea.

Legumen forma Viciæ, erectiuſculum, oblongum, compreſſum ſutura craſſiore, arcuatum, ſtyli baſi rigeſcente acuminatum, eadem qua calyx pube inſtructum ſutura magis pubeſcente, punctis numeroſis ſæpe confluentibus purpureis adſperſum, biloculare polyſpermum more generis.

Lectus cum Alchemillæ ſpeciebus, Thymis, Helianthemo et aliis in editis ſcocis circa Kurt-Bulak. Floret Majo, Junio. ♃

Obſ. Polygala Camerarii Epit. 929, quæ ab autoribus ad Aſtragalum monſpeſſulanam trahitur, totum noſtri habitum exprimit, quamvis leguminum anguſtia et longitudine videatur diverſa. Nomen triviale à maculâ carinæ et à leguminibus punctis rubris adſperſis deſumſimus.

LX. MEDI-

LX. MEDICAGO GLUTINOSA. Nov.

M. pedunculis racemofis, leguminibus reniformi-contortis calycibufque villofo-vifcidis, caulibus fubereĉtis pubefcentibus.

Planta perennis, ftatura et modo crefcendi Medic. fativæ affinis.

Caules e radice aliquot, in loco natali ordinario nempe in graminofis exaltatis fertilioribus ereĉti, in ficcis declinati, levitur pubefcentes.

Stipulæ liniari-fubulatæ; inferiores bafi denticulatæ, fuperiores integræ.

Foliola inferiora ovata, fuperiora cuneiformia, argute-ferrata margine bafeos integro.

Racemi axillares pedunculati; pedunculis fermis, reĉtis, ftriatis, circiter fefquicuncialibus.

Flores formæ communis in hoc genere, Medicaginis fativæ floribus paullo majores.

Calyx pedicellique pilis ereĉtis apice liquorem vifcidum fecernentibus obfiti.

Corolla pallida, mox in ochroleucum vergens, colore Loti filiquofæ, vexilli medio et carinæ difco faturatioribus; mox dilute cærulefcens, vexillo obfcurius lineato, carinæ autem difco conftanter luteo.

Legumina in figuram reniformem contorta, integra, villofo-vifcida ficut calyces.

Occurrit in montium graminofis fertilioribus, circa Kurt-Bulak frequens; nec in lapidofis ficcis plane exulat. Floret Junio. ♃

Obf. Medicagini fativæ affinis, vifcofitate calycum leguminumque, nec non florum colore conftanter pallidiore dignofcenda.

LXI. SCORZONERA ERIOSPERMA. Nov.

S. foliis lineari-fubulatis canaliculatis cauleque bafi fublanatis, feminibus tomento candidiffimo veftitis.

Scorzonera cretica anguſtifolia, ſemine tomento candidiſſimo. Tournef. Cor. p. 36.

Radix formæ communis in hoc genere.

Caules multi, adſcendentes, ſimplices, baſi folioſi ſublanati, ſuperne nudi, glabri, uniflori.

Folia graminea, linearia, longa, acuminata, canaliculata, dorſo carinata, nervoſa, baſi eodum modo quo caulis lana vaga albiſſima inſtructa.

Flores Scorzoneræ hiſpanicæ floribus minores.

Calycinæ ſquamæ exteriores duplo breviores circiter 10; interiores 8 læves.

Corollulæ haud multæ, luteæ, extimæ ſubinde tinctura levi purpurea.

Genitalia concoloria.

Semina tota tomento ſericeo candidiſſimo tecta.

Pappus plumoſus, multiſetus.

In graminoſis altioribus, nec non in arenoſis maris Caſpici circa Tarki et Derbentum obvia. Floret Junio. ♃

LXII. CARDUUS LAPPACEUS. Nov.

C. (foliis ſeſſilibus) foliis bifariam pinnatifidis, ſpinoſis, ſupra hiſpidis, ſubtus tomentoſis; calycibus ſubſeſſilibus folioſis, ſquamis acuminatiſſimis inermibus reflexo-uncinatis.

Caules e radice perenni nonnulli, circiter tripedales, ſimplices, nudi, albo-tomentoſi, ſulcato-angulati; ramis apicis brevibus floriferis.

Folia ſeſſilia, bifariam pinnatifida; ſegmentis alternis elongatis pectinato-patentibus, lineari-lanceolatis, apice ſpina alba valida munitis; quibus interjecta ſunt ſegmenta multo minora retroflexa apice ſpinoſa et inſuper baſi utrinque ſpina ſive denticulo ſpinoſo munitæ. Folia ſuperiora minora et minus profundè ſecta, ſpinis tamen haud minus validis: omnia ſuperne veridia pilis rigidis erectis hiſpida, ſubtus albo-tomentoſa.

Flores ad caulis ramorumque apices nonnulli approximati, ſubgloboſi, magnitudine florum Arctii Lappæ, pedunculis

breviſſimis

breviſſimis foliofis inſiſtentes; ita quidem, ut folia floralia caulinis ſummis ſimilia involucri ſpeciem referant.

Calyces glabri, virides; ſquamis linearibus, in acumen ſetaceum longum tenuiſſimum ſubinerme apice patulum deorſum arcuatum et purpuraſcens productis.

Floſculi purpurei colore et forma Cardui lanceolati.

Pappus plumoſus.

Occurrit hinc inde in montibus altioribus. Floret Auguſto. ♂

Obſ. Carduus ſcaber. Poiret. Voy. en Barbarie II. p. 231. noſtro certè affinis eſt et fortè idem. Repugnat tamen ſtatura gigantea 14-15. pedum.

LXIII. CARDUUS strigosus. Nov.

C. (foliis ſeſſilibus) foliis ſinuato-pinnatifidis ſpinoſis, lobis alternis minoribus divaricatis, ſupra trigoſis, ſubtus glabris; calycibus ovatis glabris, ſpinis ſimplicibus patulis.

Radix biennis. Caules plerumque nonnulli bipedales. tripedales et altiores, ſimplices, ſulcati, glabri; ramis floriferis verſus apicem haud elongatis, attamen longioribus quam antecedentis.

Folia utrinque viridia; ſupernè et margine pilis rigidiſſimis quaſi ſpinulis maxime hiſpida; ſubtus glabra, venis reticulata; cæterum ſeſſilia, bifariam ſinuato-pinnatifida; ſegmentis brevioribus quam antecedentis, ſed in ſpinas haud mingus validas deſinentibus; alternis breviſſimis retrorſum diſcedentibus et utrinque ſpinulâ laterali munitis ſicut antecedentis.

Flores antecedentis floribus majores, at Cardui lanceolati floribus minores.

Calyces ovati, glabri, virides, ſpinoſi; ſpinis ſimplicibus patulis rectis albicantibus.

Corollæ purpureæ forma præcedentis.

Pappus plumoſus.

E. frequentiſſimis Cardui ſpeciebus ubique in ruderatis campis et agris incultis occurrit, Floret Auguſto, Septembri. ♂

LXIV. CARTHAMUS CYNAROIDES. Nov.

C. foliis utrinque tomentofis, radicalibus pinnatafidis, caulinis oblongis adnato-decurrentibus dentato-fpinofis; calycibus folitariis aphyllis, fquamis apice triangularibus reflexis.

Radix fimplex, defcendens, biennis.

Caulis erectus, circiter pedalis, raro bipedalem altitudinem attingens, bafi fimplex, fuperne divifus in ramos 2 aut 3 unifloros, lana alba molliffima tenui veftitus.

Folia utrinque lanata ficut caulis; radicalia et caulina inferiora lyrato-pinnatafida, dentata, dentibus fpinofis divaricatis; fuperiora oblonga non diffecta, fed tantum dentata; caulina omnia nervo medio cauli adnata fic decurrentia, apice foluta.

Pedunculi breviffimi, nudi, arachnoideo-villofi ficut calyces.

Calyx globofus, magnitudine Cardui Eriophori; fquamis exterioribus apice triangulari reflexo carinato, fpina terminali et quibufdam minutis lateralibus; fquamis interioribus fubulatis, membranaceis, pungentibus, erectis.

Corollulæ candidæ; tubo lineari; limbo paullo craffiore, bafi monophyllo cylindrico, laciniis linearibus.

Genitalia concoloria. Antheræ corollulas notabiliter excedentes. Stylus bifidus antheras vix fupereminens.

Receptaculum villis longis rigidis tortis inftructum.

Pappus fetaceis, receptaculi villis fimilis, fugax.

Semina angulata, coftis elevatis nigro-punctulatis; grifea punctifque minimis excavatis infculpta.

Crefcit paffim in ruderatis, ad vias et aggeres inter Kubam et Schamachiam. Floret Junio, Julio. ♂

Obf. Adoptata complurium è recentioribus methodo hanc plantam ob calycinas fquamas appendiculatas duplicato-fpinofas ad Carthamum retuli. Nomen triviale analogiam imprimis florum majufculorum calycibus globofis duris præditorum, cum Cynara indicat.

LXV. CARTHAMUS GLAUCUS. Nov.

C. ramis ſubcorymboſis paucis, foliis ſuperioribus villoſo-viſcidis glaucis, ſpinis foliaceo-marginatis, calycibus, folioſis ſquamis intimis lanceolatis æqualibus.

Cnicus Atractylidis folio et facie incanus patulus flore purpuraſcente. Tournef. Corr. p. 33.

Radix annua ſimplex.

Caulis circiter pedalis erectus læviuſculus, baſi ſimplex; ſuperne ramis paucis ſubunifloris, ſubcorymboſis, attamen haud raro quoque ipſo caule altioribus.

Folia dentato-pinnatafida ſpinoſa; radicalia profunde ſecta fere pinnata, ſegmentis oppoſitis anguſtis patentibus ſpinoſis margine ſpinuloſis, impari majore ovato-indiviſo: caulina minus lacera, ſuperiora imprimis floralia pube tenui ſubviſcida glauceſcentia immo in cæſium vergentia.

Flores ad apicem caulis et ramorum ſolitarii, oblongi, foliis ſupremis caulynis ſimilibus involucrati.

Squamæ calycinæ propriæ ſubmembranaceæ, apice ſcarioſæ, omnes etiam intimæ lanceolatæ æquales et in mucronem inermem excurrentes.

Corollulæ purpureo-cæruleæ colore florum ſerratulæ arvenſis; tubo filiformi longo fauce inflata lineis quinque obſcurioribus laciniis filiformibus.

Antheræ corollæ concolores eademque paullo breviores. Filamenta ſub apice villoſa.

Stylus ſimplex corollula longior. Stigma ſimplex.

Receptaculum conicum ſetoſum.

Semina exteriora pappo deſtituta; cæterorum pappus ex ſetis numeroſis membranaceis læviſſimis gilvis conſtat.

Occurrit rarius in collibus apricis, lectus in vicinia rivulorum Giljeni et Atæ. Floret Junio, Julio. ☉

LXVI. CARTHAMUS LANATUS.

C. ramis ſubcorymboſis paucis, foliis glabriuſculis viridibus ſpinis foliaceo-marginatis, calycibus folioſis, ſquamis intimis ſub apice dilatatis ſerrulatis.

Carthamus lanatus: caule pilofo fuperne lanato, foliis inferioribus pinnatafidis, fummis amplexicaulibus dentatis. Linn. ed Reich. 3. p. 698.

Copiofus ubique in pafcuis, agris, ad vias et aggeres. Floret Junio, Julio. ⨀

Obf. Præcedenti maximè affinis, nec ulla alia nota, quàm defectu pubis fubvifcidæ, fquamarum calycinarum intimarum forma, denique corollulis ochroleucis difcrepans. In Tauriâ quoque occurrit, ubi Carthamus glaucus cum infequenti fpecie exulat.

LXVII. CARTHAMUS oxyacantha. Nov.

C. ramis corymbofis pluribus, foliis glabriufculis lucidis fpinis tenuiffimis coloratis, calycibus foliofis.

An Cnicus orientalis humilior flore flavo. Tournef. Corr. p. 33?

Radix annua, formæ præcedentium.

Caulis plerumque pede brevior, bafi fimplex, fuperne in ramos floriferos numerofiores patentiores et habitiorem corymbum efficientes fubdivifus, albiffimus, fuperne villofus.

Folia minus profunde fectâ dentibufque minoribus quam antecedentium apice tantum nec unquam margine fpinofis; cæterum læte viridia cum nitore, et quamvis leviter pubefcant nequaquam etiam fumma glaucefcentia. Spinæ crebiores, flavæ, exiles et nudæ nec ut antecedentium margine foliaceo ipfum tantum apicem liberum relinquente cinctæ, fed ftatim ab origine e folii fubftantia liberæ.

Flores forma et involucro præcedentium.

Flofculi pauciores, toti faturatius lutei, lineis nullis picti; limbo æquali bafi non campanulato ut in præcedentibus, fed ad ipfum tubum ufque fiffo laciniis linearibus.

Genitalia corollulis concolaria.

Lectus inter fegetes in vicinia rivulorum Atæ et Giljeni nec alibi. Floret Junio. ⨀

Obf. Hæc quoque fpecies plurimis notis cum præcedentibus convenit, facilius tamen ab his diftinguitur, quàm hæ inter

inter ſeſe invicem. Caracteres diagnoſticos majoris momenti ipſa ſpeciei definitio exhibet.

LXVIII. CARLINA Echinus. Nov.

C. foliis ſpinoſis ſinuatis dentatis ſubtus tomentoſis; calycinis ſquamis ciliatis floſculis brevioribus, exterioribus ſpinoſis patulis, interioribus ſubulatis reflexis.

Radix biennis?

Caulis ſeſquipedalis aut bipedalis, erectus ſuperne ramoſus, ſulcato-angulatus mediante nervo medio foliorum decurrente, ſubtomentoſus.

Folia ſemiamplexicaulia, oblonga, ſinuata; lobis dentatis, dentibus divaricatis apice ſpina tenui longa et margine ſpinulis nonnullis parvis armatis; ſupra leviter pubeſcentia, ſubtus tomento tenui caneſcentia; ſuprema minora, ſpinis crebris bifariam patentibus longiſſimis inſtructa.

Flores in ramis ramuliſque terminales, ſubſeſſiles ſolitarii, nutantes, circiter Jaceæ magnitudine.

Calyx ſubgloboſus, imbricatus; ſquamis oblongis, exterioribus in ſpinam longam baſi ciliatam albidam patulam deſinentibus, interioribus membranaceis anguſtis ſubulatis ciliatis vix pungentibus recurvis.

Floſculi omnes æquales hermaphroditi ſulphurei; tubo filiformi tenui, limbo erecto, laciniis filiformibus.

Genitalia concoloria vix corollularum longitudine. Stylus apice vix fiſſus.

Receptaculum piloſo-paleaceum, nempe pilis baſi in membranulam breviſſimam connatis.

Semina albida, lævia, forma ſeminum Cardui; pappo plumoſo fere floſculorum longitudine.

Hinc inde obvia in lapidoſis ſiccis, imprimis ad rivorum margines in tractu montium inter Kubam et Schamachiam. Floret Julio. ♂

Obſ. planta hæc æque ac ſpecies affines Carlina atractyloides Linn. et Carl. gorterioides Lamarkii ægre ad hoc genus refertur; a Carduo quoque et à Carthamo aliena.

LXIX. GNAPHALIUM CANDIDISSIMUM. Nov.

G. fublignefcens, niveo-tomentofum; foliis ellyptico-lanceolatis fucculentis; corymbo compofito; calycibus ovatis pubefcentibus.

Elichryfum orientali fimile, calyce florum argenteo. Tournef. Corr. p. 33?

Caules e bafi lignefcente plures, dodrantales aut breviores, declinati, fimplices, florum corymbo terminati, tomento candidiffimo denfo ceu epidermide obducti æque ac pedunculi et folia.

Folia fparfa, ellyptico-lanceolata, brevia, obtufiufcula, integerrima, fucculenta; fuperiora minus conferta fenfimque minora et anguftiora.

Flores in corymbos denfos multifloros difpofiti, ovati, Gnaphalii arenarii floribus duplo majores.

Calycinæ fquamæ imbricatæ, obtufæ; exteriores breviores in medio tomento tenui albo veftitæ margine dilutiffimè fufcefcentes; interiores longiores albo-membranaceæ, linea dorfali apicem non attingente viridi.

Flofculi omnes æquales hermaphroditi; limbo 5-fido luteo.

Antheræ luteæ, corollulas vix fupereminentes.

Styli in florente vix ftaminum longitudine.

Receptaculum nudum.

Semina minutiffima, pappo pilofo multifeto fere longitudine flofculorum.

Commune ad margines lapidofos rivulorum imprimis rivi Pirfagat in vicinia veteris Schamachiæ. Floret Julio, Augufto. ♄ ♃

Obf. Videtur Gnaphalio orientali non diffine, foliis brevibus ellyptico-lanceolatis fucculentis et tomento denfo candidiffimo diftinguendum.

LXX. ANTHEMIS FRUTICULOSA. Nov.

A. fuffruticofa diffufa incana, ramis adfcendentibus fuperne nudis unifloris, foliolis linearibus carnofis, petalis ovatis.

Anthemis

Anthemis fruticofa, foliolis linearibus carnofis, pedunculis longioribus nudis erectis unifloris. Gerard. Prov. p. 209. tab. 8.

Chamæmelum montanum incanum abfynthioides. Barrel. ic. 451.

Radix multiceps, fibris longiffimis.

Caulis bafi fruticulofus diffufus, emittens fafciculos foliorum et ramos floriferos erectos, fpithamæos fimplices, fuperne nudos unifloros, villis incumbentibus albis pubefcentes ficut tota planta.

Folia petiolata, pinnato-compofita; pinnulis alternis oblongis brevibus carnofis, apice mucrone minimo.

Pedunculi minus quam caulis canefcentes, fubtiliffime ftriati.

Flores magnitudine Cotulæ.

Calyx hemifphæricus albo-villofus; fquamis ovatis margine tenuiffime membranaceis, apice appendiculo fcariofo lacero dilute-frufcefcente; interiorum magno dilatato.

Corollulæ radii albæ, latæ et breves, apice crenulis 3 aut 4, tubo carnofo compreffo.

Corollulæ difci 5-fidæ, luteæ, tubo carnofo.

Stamina radii filamenta 5 antheris carentia.

Stamina difci fertilia vix corollulis longiora.

Stylus bifidus; radii vix tubum corollulæ fupereminens; difci fere corolluleæ longitudine.

Receptaculum hemifphæricum, paleaceum; paleis membranaceis, linea media viridi in acumen elongata, fere longitudine tubi corollularum.

Semina nuda, angulata, pappo deftituta.

Copiofa in abruptis fteriliffimis circa Kurt-Bulak. Floret Junio. ♄

Obf. Anthemidi alpinæ Linn. affinis, fed præter alias notas caulibus fuffruticofis ramofis diffufis difcrepans.

LXXI. ACHILEA Eupatorium. Nov.

A. villofa; foliis pinnatis; pinnulis plano patentibus, incifo-ferratis, bafi latiore decurrentibus, corymbis convexis ampliffimis, floribus radio deftitutis.

Ptarmica

Ptarmica orientalis foliis Tanaceti incanis flore aureo. Buxb. C. 2. p. 25. tab. 19.

Ptarmica orientalis incana, foliis pinnatis, semiflosculis florum vix conspicuis. Tournef. Corr. p. 38.

Radix perennis.

Caules sæpe nonnulli, erecti, pedales, simplices, sulcati, villosi et subcanescentes sicut tota planta neque vero albo-tomentosi.

Folia pinnato-composita, lata, plana, sessilia; foliolis oblongis, acutiusculis, inciso-serratis, basi latioribus et postice ad petiolum communem decurrentibus adeoque confluentibus.

Corymbus terminalis amplissimus, convexus, densus e floribus numerosissimis constans.

Flores oblongi, magnitudine. Ach. Millefolii.

Calyx villosus; squamis oblongis, quibusdam ad basin brevioribus.

Radius nullus, sed flosculi omnes tubulosi, circiter deni, lutei, genitalibus concoloribus.

Receptaculum paleaceum; paleis oblongis albo-membranaceis, nervo medio viridi; ætate elongandum in rachis speciem, cui flosculi cum paleis adhærent.

Semina formæ communis in hoc genere, minutissima, pappo carentia.

Odor fortis generi Achilleæ peculiaris.

Communis ad vias et sylvarum margines, per totam regionem circa Derbentum, Baku, Kubam, Schamachiam utramque. Floret Junio. ♃

Obs. Synonymon Buxbaumii certe huc pertinet, quod locus quoque natalis ab eo citatus testatur. Cæterum species nostra proxima videtur Achilleæ filipendulinæ. Lamarkii; conf. ej. Dict. Enc. Bot. 1. p. 27. n. 12.

LXXII. JUNIPERUS EXCELSA. Nov.

J. caule arboreo validissimo; adultæ foliis minimis oppositis quadrifariam imbricatis; tenellæ acerosis patulis ternis; baccis nigris.

Cedrus

Cedrus orientalis fœtidiſſima arbor excelſa; ſeu Sabina orientalis fructu parvo nigro. Tournef. Corr. p. 41.

Juniperus Sabina var. Taurica. Pallas Roſs. 2. p. 15.

Juniperus hiſpanica: foliis quadrifariam imbricatis acutis. Mill. Dict. n. 13?

Communis in montium altiorum lapidoſis aridis. ♄

Obſ. Differt à Junipero Sabinâ cui proxima proceritate et mole trunci. Folia adultioris vix à Sabinæ foliis diſtinguenda; at tenella folia gerit planè abſimilia, ſcilicet acероſa, pungentia, patula, terna, Juniperi communis foliis analoga. Deſcriptionem in Florâ Tauricâ dabimus.

LXXIII. SACCHARUM Ravennæ. Linn.

A. panicula elongata contracta, juniore ſpiciformi; pedunculorum calycumque lana longiſſima floſculos ſubmuticos abſcondente.

Radix repens, amplum ſæpe ceſpitem foliorum culmorumque promens.

Culmi ſeſquipidales, bipedales, immo quandoque quadripedales, teretes, ſuperne nudiuſculi.

Folia lata, arundinacea, acuminata, læte viridia, vaginis ore piloſis.

Panicula terminalis elongata, contracta; junior immo dum floret ſpiciformis et Alopecuro analoga; ætate patentior.

Paniculæ rami alterni, imbricati, ſubdiviſi, lana candida longiſſima et molli veſtiti; quæ lana.

Spiculas, parvas, lineares, alternas, approximitas breviter pedicellatas abſcondit.

Floſculi plerique feminei; maſculi rari.

Calyx gluma uniflora bivalvis; valvulis oblongis membranaceis albis, dorſo ſpadiceis ſubæqualibus muticis, totis lana eadem qua paniculæ rami pedicellique obſitis.

Corolla univalvis; valvula minima albomembranacea, ſubmutica, calyce involuta.

Stamina ſpicularum maſcularum tria. Antheræ flavæ admodum fugaces.

Germen

Germen ſpicularum fœminearum oblongum; ſtyli duo, calyce duplo longiores, plumoſi, purpurei.

Occurrit in planitie verſus torrentes Keſſartſchai et Samur inter Kubam et Derbentum. Floret Junio. ♃

LXXIV. MIMOSA Stephaniana. Nov.

M. (aculeata, foliis bipinnatis) pubeſcens, foliolis ovato-oblongis baſi inæqualibus, ſpicis ſimplicibus elongatis, ſtaminibus vix corolla longioribus.

Senna Perſica ſpinoſa, foliis exiguis. Buxb. C. 3. p. 36. tab. 48. Mala.

Siliqua Nabathæa ex Perſia allata. Breyn. Exot. Cent. 1. tab. 55. legumen optime.

Radix lignoſa, craſſitie digiti, extus aſtro-purpurea, intus albida, longiſſime in terram defixa.

Caulis fruticoſus, erectus, pedalis aut bipedalis, raro altior, craſſitie pennæ anſerinæ, aculeatus aculeis ſparſis rectis baſi dilatatis, pube ſubtili glaucеſcens.

Rami elongati, graciles, alterni, adſcendentes flexuoſi, aculeati et pubeſcentes ſicut caulis præſertim juniores.

Stipulæ ſolitariæ, caducæ; inferiores ſubreniformes obſolete denticulatæ; ſuperiores minores ſcilicet vix foliola magnitudine ſuperantes, integræ, oblongæ, acutæ.

Folia abruptè bipinnata, circumſcriptione ovata, breviter petiolata, pubeſcentia ſicut caulis; pinnis oppoſitis conjugationibus ſæpe quinis; petiolo communi in mucronem muticum breviſſimum excurrente.

Foliola ſubtredecemjuga, ſtricte oppoſita, ima et ſuperiora paullo minora; omnia parva, oblonga, acutiuſcula, obliqua et ferè ſemicordata ob baſin exteriorem folioli productam nervumque principalem à baſi ad apicem non per folioli medium ſed margini antico proprius excurrentem. Juxta nervum principalem alter ex ipſâ baſi marginem interiorem verſus exit, qui vero folioli apicem haud attingit.

Racemi axillares ſimpliciſſimi, ſolitarii, folio longiores, breviter pedunculati.

Flores

Flores circiter magnitudine florum Veronicæ, pallide lutescentes, breviſſimis tenuiſſimiſque pedicellis inſiſtentes, ſparſi, bracteis vix ullis.

Calyx minimus, denticulis quinque.

Corolla regularis, pentapetala; petalis oblongis, acutiuſculis, parum patentibus.

Stamina 10 receptaculo inſerta, corollo paullo longiora. Filamenta ſubulata, baſi vix connexa equalia. Antheræ incumbentes flavæ.

Piſtillum unicum ſtaminibus paullo longius; germine oblongo; ſtylo columnari; ſtigmate ſimplice, torto.

Legumen oblongum, circiter ſeſquiunciale, curvum, obtuſum, farctum, purpuraſcens, multiloculare, ſcilicet duplici ſeminum ordine intra materiam ſuberoſam; ſingulis diſſepimento tranſverſali ſeparatis.

Semina formæ communis in hoc genere, magna, compreſſa, ſubrotunda, læviſſima, pallide fuſca.

Flores mere maſculos nullos vidi; attamen plurimi e floribus abortiunt, ita ut raro legumina tria quatuorve in eadem ſpica matura ſemina ferant. Folia autumno decidunt in ſpecie hac frigorum magis quam congeneres omnes patiente.

Frequens in ſiccis provinciæ Schirvan, imprimis per planitiem ad Cyrum Fluviam, circa novam Schamachiam et in viciniâ fluminis. In deſertis ad dextrum latus Cyri, Araxem verſus, integras plagas ferme ſola occupat. Floret Julio. ♄

Dicta in honorem D. Friderici Stephan, Profeſſoris Moſquenſis.

Abſolutis

Absolutis stirpibus novis vel minus cognitis, nonnullarum, quarum hactenus mentionem haud feci, nomina subnecto.

ZIZIPHORA capitata.
Salvia viridis.
.... verbenaca.
.... verticillata.
.... ſylveſtris.
.... Æthiopis.
.... Sclarea.
Iris pumila.
.... ochroleuca.
Saccharum Ravennæ.
Phalaris bulboſa.
Panicum dactylon.
Chrypſis phleoides (Phleum Schœnoides Linn.)
.... ſchœnoides (Schœnus aculeatus Linn.)
Alopecurus monſpelienſis.
Milium paradoxum.
Agroſtis bromoides.
Melica ciliata.
Poa ſalina.
Briza virens.
Cynoſurus echinatus.
Stipa pennata.
Avena fatua.
Arundo Donax.
.... Phragmitis.
Arundo Calamagroſtis.
.... epigejos.
Triticum junceum.
.... proſtratum.
Dipſacus laciniatus.
Scabioſa tranſylvanica.
Scabioſa leucantha.
Knautia plumoſa.
Galium rubioides.
.... pariſienſe.
Plantago Pſyllium.
.... Loeflingii.
Camphoroſma monſpelienſis.
Anchuſa italica.
Aſperugo procumbens.
Echium rubrum.
Meſſerſchmidia Arguzia.
Androſace maxima.
Convolvulus lineatus.
.... Cantabrica.
.... Perſica.
Verbaſcum phlomoides.
Cynanchum acutum.
Apocynum Venetum.
Chenopodium Botrys.
Beta maritima.
Velezia rigida.
Tordylium maximum.
Caucalis orientalis.
Daucus Viſnaga.
Ferula nodiflora.
Laſerpitium aquilegifolium.
.... ferulaceum.
.... prutenicum.
Statice coriaria Pall.
.... fruticuloſa.
Linum auſtriacum.
.... flavum.
.... gallicum.

Allium

Allium ſphærocephalum.
.... deſcendens.
.... moſchatum.
.... flavum.
Ornithogalum Narbonenſe.
Bulbocodium vernum.
Atraphaxis ſpinoſa.
Rumex ægyptiacus.
.... nemolapatho Ehrh. affinis.
Epilobium roſmarinifolium. Curtis.
Stellera Paſſerina.
Zygophyllum Fabago.
Gypſophila paniculata.
.... muralis.
.... glomerate Pall. affinis.
Dianthus prolifer.
.... deltoides?
Dianthus Carthuſianorum.
Cucubalus italicus.
.... molliſſimus.
Silene conoidea.
...... chloræfolia Smith.
Arenaria faſciculata Jacq.
.... ſtriata Allionii.
Garidella Nigellaſtrum.
Peganum Harmala.
Euphorbia ſaxatilis.
.... gerardiana Jacq.
Potentilla ſupina.
.... recta.
Chelidonium corniculatum.
Ciſtus fumana.
...... Ægyptiacus.
...... helianthemum. Lamarkii.

Ciſtus niloticus.
Nigella orientalis.
Clematis orientalis.
Thalictrum minus.
Teucrium hircanicum.
.... Polium.
Nepeta ucrainica.
Sideritis montana.
Stachys lanata.
.... recta.
Marrubium peregrinum.
Phlomis laciniata.
.... Herba Venti.
.... tuberoſa.
Thymus vulgaris.
.... Zygis.
.... acinos.
Scutellaria orientalis.
Prunella laciniata.
Rhinanthus Trixago.
Lathræa Phelypæa.
Pedicularis tuberoſa.
Dodartia orientalis.
Myagrum rugoſum.
Anaſtatica ſyriaca.
Lepidium latifolium.
.... veſicarium.
Turritis hirſuta.
Geranium ciconium.
Althæa hirſuta.
Ononis minutiſſima.
Lathyrus ſphæricus Retrii?
.... hirſutus.
.... piſiformis.
Vicia Narbonenſis.
Coronilla Securidaca.
Hippocrepis uniſiliqua.

Hedyſarum

Hedyſarum Buxbaumianum Pall.
.... grandiflorum.
.... Alhagi.
.... cornutum.
Phaca ſalſula Pall.
Aſtragalus auſtriacus.
.... exſcapus. var. β.
.... trimeſtris.
Trifolium lappaceum.
.... anguſtifolium.
.... fragiferum.
Lotus Dorycnium.
Hypericum Coris.
Tragopogon major Jacq.
.... undulatus. —
Scorzonera hiſpida Forſk.
.... dichotoma.
Vahlii.
Picris echioides.
Chondrilla juncea.
Crepis neglecta.
Hyoſeris hedypnois.
Serratula amara.
.... centauroides.
Carduus polyanthemus.
.... ciliatus.
.... marianus.
.... hiſpanicus Lamark.
.... mollis. Jacq
Carthamus Acarna Lamark. (Cnicus Linn.
Chryſocoma villoſia.
.... Linoſyris.
Artemiſia pauciflora.
.... humilis.
.... rupeſtris.
.... piperita, Pall.
Gnaphalium arenarium.
.... orientale.
Xeranthemum annuum.
Conyza ſquarroſa.
Erigeron acre.
.... canadenſe.
Senecio erucifolius.
Cineraria longifolia?
Inula Oculus Chriſti.
..... germanica.
Aſter Amellus.
Chryſanthemum inodorum.
.... alpinum.
Achillea tomentoſa.
Buphalmum ſpinoſum.
Centaurea crupina.
.... acuta. Vahlii.
.... ſibirica.
Centaurea ſplendens.
.... amara.
.... ſolſtitialis.
.... reflexa.
Echinopſis ſtrgoſus.
Serapias latifolia.
Andropogon Iſchæmum.
Ægilops ſquarroſa.
.... triuncialis.
Atriplex hortenſis. var.
.... laciniata.

NOTES

TO VOL. I.

PAGE 1. The Volga, in many points of view, is amongſt the moſt remarkable rivers of the Ruſſian empire. After a courſe of above 4000 verſts, through eight governments, it falls into the Caſpian near Aſtrachan, through more than ſixty-five branches. It is navigable up to Twer, and is increaſed by other conſiderable rivers, of which the Kama and Oka are the moſt notable, as they form a communication between the moſt fertile provinces. The weſtern branch is ninety verſts long, the eaſtern thirty only. Moſt of the mouths are ſo much choked up with ſand, that, of the ſeven large ones, but two can be navigated by veſſels of 150 tons. Thoſe which ſail from Aſtrachan in ſummer or autumn, frequently require ſome weeks to reach Tſchetire Bugri, (the four hills,) one of the outermoſt iſlands, that ſerves as a landmark to fetch into the channel. Veſſels, going out and in, avail themſelves as much as poſſible of the ſpring inundations, when the above paſſage may be made in two days; yet an accurate knowledge of the channel is requiſite, to avoid the ſhallows. All veſſels, ſailing up or down, muſt lie-to at Tſchetire Bugri, to be viſited by the cuſtom-houſe and health-office. *Storch.*

IBID. The Don *(Tanais)* takes its riſe from the lake Ivan, in the government of Toola; and paſſing through the whole province of Aſov, divides itſelf near Tſcherkaſk into three ſtreams, and falls into the ſea of Aſov. It extends its courſe about 1000 verſts, and has ſo many windings, is in many parts ſo ſhallow, and abounds with ſuch numerous ſhoals, as to be ſcarcely navigable, excepting in ſpring upon the melting of the ſnows; and its mouth is alſo ſo choked with ſand, that flat-bottomed veſſels only can paſs into the ſea of Aſov at any other ſeaſon. The banks of the Don, and of the rivers which fall into it, are clothed with large tracts of foreſts, whoſe timber is floated down to St. Demetri and Roſtov, where the frigates for the ſea of Aſov are chiefly conſtructed. The navigation, Mr. Coxe obſerves, may poſſibly hereafter be rendered highly valuable, by conveying to the Black Sea the iron of Siberia, Chineſe goods, and Perſian merchandiſe; which latter commodities, as well as the products of India, formerly found their way into Europe through this ſame channel. In the time of Peter the Great, ſays Pleſcheef, a project was formed to cut a channel from the river Ylavlya to the ſource of the Kamyſhinka, for the ſake of making a communication between the Don and the Volga; but the ſcheme was abandoned, becauſe the current of the former was found to be higher, near Sarepta, than that of the Volga. The Don divides Europe from Aſia, and frequent mention of it is made by ancient writers.

Ibid. On the crown-lands, near Zarizin, are ſprings of mineral water known by the name of Sarepta, which in the ſummer ſeaſon draw a vaſt concourſe of people. In Pallas's Travels the reader will find a particular account of the colony of Herrnhüter at Sarepta. This ſect aroſe from ſome emigrants from Moravia, who forſook their poſſeſſions on account of the perſecutions they ſuffered, as Proteſtants, from the Roman Catholics. They retired to an eſtate belonging to the family of Count Zinzendorf, about fifty miles E. of Dreſden. It is ſituated upon the riſe of a hill called *Hutberg*, or Watch-hill, from which they took occaſion to call their new ſettlement *Herrnhut*, or the "Watch of the Lord." The building was begun in 1727. The increaſe, and admirable regulations of this ſettlement, occaſioned no ſmall ſurpriſe in the neighbourhood; and cauſed, in the years 1732, 1736, and 1737, commiſſioners to be appointed to examine into the doctrines and proceedings of the brethren. They made a favourable report; and ever ſince both Herrnhut,

hut, and other ſettlements of the united Brethren in Saxony, have been protected; and even ſeveral immunities offered them by the court, but not accepted. The colony was viſited in 1766, by Joſeph II. and by Fred. William II. king of Pruſſia, and ſeveral other royal perſonages, who expreſſed their ſatisfaction on examining its peculiar regulations. They have ſettlements in many parts of the world: in England, the principal ones are at Fulnee near Leeds, and at Fairfield near Mancheſter; they have likewiſe a chapel in London. Perhaps there is no denomination in whom a meek, quiet, child-like ſpirit, has been more cultivated.

Ibid. *Steppe*, or *Stept*, is the common word in Ruſſia for deſert. Dutens ſtates the *verst* to be 500 *sazen*, or fathoms, each compoſed of three ells of Ruſſia, or ſeven feet Engliſh: ſo that the verſt contains 1166 yards 2 feet. A German *meile* contains ſomething leſs than five Engliſh miles.

Ibid. Pliny ſays (lib. vi. cap. xix.) that the Scythian called Caucaſus, *Groukasus*, *i. e. nive candidum*, becauſe i was covered with eternal ſnow, but that is not the caſe at preſent.—*Schrœder.* The learned Bochart conjectures, that Caucaſus took its name from Gog, as inhabiting it, or the parts about it: for he obſerves, that *Gog-chasan* denotes, in the neighbouring oriental tongues, as much as Gog's Fort, and from Gog-chaſan the Greeks framed the name Καύκασος.

Ibid. The words Galbus and Jalbus are probably of Phœnician, or Celtic origin. *Alp*, or *Alb*, in Phœnician, denotes *high* mountain; *Alben*, in Hebrew, ſignifies *white*. The word *Alp* implied, among the ancient Scythians and Scandinavians, both a mountain and a mountain ſpirit; it being a perſuaſion among them, that mountains and rocks were inhabited by dæmons. The iſland of Britain was formerly called *Albion*. See the Cyclopædia.

Page 8. Major Rennell ſtates, that Ecbatana, a ſplendid city in India, the ſummer capital of the Perſian monarchs, was unquestionably on, or near the site of modern Hamadan in Al Jebal. See the Scripture account of it in Judith, c. i. It is said to be the burial-place of Queen Esther and Mordecai. When Mr. Hanway was there with Nadir Shah, it was his grand repository for cannon, bombs, and other artillery.

PAGE 13. The Manitsch, according to Pallas, rises 180 versts from the Saipa, in a low plain more than twenty versts wide. It flows eastward through a vast bog. It then runs 100 versts through an immense arid plain, which is said to extend to the Don and the Steppes of the Taurid. At some distance from its source, it forms two salt lakes, called the "Holy Lakes." It presents traces of the waters having flowed much more precipitately towards the Black Sea. Its current is very slow, and the water very salt. Pallas reckons twenty-four hours journey from the source of the Manitsch to the Kuma, which flows from W. to E. and is very rapid.

PAGE 16. The following altitudes of the most considerable mountains may not be unacceptable. They are taken from different travellers; but we do not answer for their correctness.

	Yards.		Yards.
Cayambe Orcon in the Cordilleras	6060	M. Lomnitz in Hungary	2880
M. Ætna	5000	St. Bernard	2798
M. Blanc	4900	Wharaarai in Owyhee	2666
M. Rose	4780	M. Vellino	2624
Peak of Argentière	4188	M. Krivan in Hungary	2606
M. Ophyr in Sumatra	4001	Stange Alpe in Styria	2380
M. Ortele in the Tyrol	4000	St. Gothard	2300
Peak of Teneriffe	3808	Budislau in Transilvania	2296
Peak of Ossian in the Pyrenees	3500	M. Rotondo	2098
Hochorn in Salzburg	3544	M. d'Or	2096
M. Salasses, a volcano in the isle of Bourbon	3200	Simplon	2024
		Kchoes in Caucasus	1788
		Ellborus	1550
Terglau, the highest of the Julian Alps	3031	M. Sila in Calabria	1544
		M. Brenner in the Tyrol	1493
M. Lebanon	3000	Ben Nevis	1462

Vesuvius

Veſuvius	1300	Buchberg, the higheſt baſaltic mountain in Germany	914
Snowdon	1185		
M. Athos in Macedonia	1100		
Radicofani	1020	M. Cenis	868

PAGE 20. The Terek, in October 1785, cauſed a violent inundation. The road along its banks, built on arches at a very great expence, was entirely deſtroyed and ſwept away; and the neighbourhood of Kiſlar was completely under water. *Schrœder.*

PAGE 30. If ever the noble collection of paintings belonging to Prince Potemkin, or that of Count Stroganoff, ſhould be made public, no one will be aſtoniſhed at learning that Mount Ararat ſmokes and vomits fire. Reineggs and his travelling companion ſaw it in that ſtate for three days together. The oldeſt inhabitants do not recollect, nor does the Armenian hiſtory mention, any thing of this mountain's emitting flames. On that account the neighbourhood was greatly alarmed. *Schrœder.*

Struys, the Dutch traveller, who aſcended M. Ararat in July, 1670, gives a curious account of it. "I found there," ſays he, "a kind of dark yellow mineral, very ponderous and ſparkling; but could never learn what metal it would have produced when melted. We were ſeven days before we reached the hermit's cell, travelling every day five miles. In our journey upwards, we paſſed through three ſorts of clouds. The hermit's cell was hewn out of the rock, where he had reſided three and twenty years." From Tournefort's account M. Ararat ſeems to conſiſt chiefly of free-ſtone, or calcareous ſand-ſtone. He attempted to aſcend it, but failed.

PAGE 34. *Derbend Name*, or the hiſtory of Derbend. The beginning and end are wanting. This fragment is under the care of the Leſghaean judges in Kubten, and attributed to the fifth century of the Hegira. I obtained permiſſion to tranſcribe it, and the copy is depoſited in the muſeum at Gœttingen. There are other hiſtories of Derbend, but they are not deemed authentic, on account of their containing insipid fables and Tatarian accounts of giants. *Reineggs.*

Page 47. An iron ball, with sharp points, (or toppus,) is supposed to be one of the first kinds of arms used in the world. For, each of the horsemen in the bas-reliefs of Persepolis, which are certainly the most ancient monuments in the universe, carries one hanging on his horse's back. *Chardin's Travels.*

Page 49. Of all religions known, that of the Parses is perhaps the only one in which fasting is neither meritorious, nor even permitted. The Parse thinks to honour Ormusd by living well, because the body, by being fresh and vigorous, fortifies the soul against evil genii; and a man, the less he feels want, can read the *word* with more attention, and is more encouraged to do good works, consequently many celestial spirits are specially commissioned to watch over the well-being of man. *Zend Avesta.*

Page 50. It is remarkable, says Mr. Clarkson, that two of the most civilized nations, as well as two of the wisest men of antiquity, should have differed in their opinions with respect to dancing. The Greeks considered it as a wise and an honourable employment; and most of the nations, therefore, under that appellation, inserted it into their system of education. The name of *Dancer* was so honourable, as to be given to some of their gods. Statues are recorded to have been erected to good dancers. Socrates is said to have admired dancing so much, as to have learnt it in his old age. Dancing, on the other hand, was but little regarded at Rome. It was not admitted even within the pale of accomplishments. It was considered at best but as a sorry and trivial employment. Cicero says, "No man dances, in private, or at any respectable entertainment, except he be drunk or mad." The Quakers, in this respect, side with the Romans. They think it unnecessary to teach their children dancing, as an

accom-

accomplishment; because they can walk and carry their persons with sufficient ease and propriety without it.

PAGE 65. This change in the course of the Terek took place in 1769. Notwithstanding the impediments to navigation occasioned by it, Kislar continues to be the emporium for supplying the small forts on the Terekian line, and of the Caucasian trade. It receives provisions and goods for exportation from Astrachan, which are partly sent to Mosdok and partly to Caucasus. The principal article imported is silk. In four years from 1767 to 1770, silk to the amount of 5966 pouds was imported. Statement of the trade as follows:

	From 1773 to 1777.	From 1793 to 1797.
Imports	96,843 roubles.	91,619 rs.
Exports	51,794	52,906
Total	148,637	144,525
Balance in favour of the former	45,049	38,713

Storch's Pict. of Russia.

PAGE 70. To give the reader an idea of the force of the Lesghaes, the following is the list of troops, (as stated by Hanway,) which they sent to the Russian General, when they implored the protection of Russia against Nadir Shah:

Achmed Khan has	12,000
The tribe of Aparz	13,000
The Khan of Schunketin	2,700
In the districts of Kanschukul	8,000
In Aby	5,000
In Abugal and Kalackski	7,000
In the districts of Kusti	500
In Kly	2 500
In Karack	7,500
In Gedat	4,000
In Kinjode	1000
In Kurada	1000
	64,200

PAGE 72. The natives of the province of Gilan call themſelves *Gelæ*, and by their neighbours they are called Gelaky. They ſpeak a peculiar language, and have the character of being honourable men; which is not the caſe, however, with the Perſians who are united with them. *Reineggs*.

PAGE 73. The Tatars write and call the Volga, *Bulga*. By the Caucaſian nations and others it is called *Ethel*, or *Ethil*. *Bulgar* implies every thing in the neighbourhood of that river, and that comes from it. *Reineggs*.

IBID. Public buildings and moſks remain ſtill in good preſervation at Mattſchar, and from dates on the grave-ſtones, and other Arabic inſcriptions, we find that the city now lies deſolate above 180 years; and this in one of the fineſt and moſt fertile countries. *Reineggs*.

Pallas ſtates, that ſome of the ſepulchral monuments are ſixteen yards in height, and large in proportion. The Kalmuks call the ruins *Zaghan-Balgazun*, the white city.

PAGE 75. The Scythian nations could as little formerly, as at preſent, pronounce words beginning with harſh conſonants, without prefixing a vowel of ſimilar ſound. Thus, inſtead of *Russ*, *Rum*, and *Styria*, they ſaid Uruſs, Urum, and Iſtiria. *Schrœder*. We have a curious ſimilar inſtance amongſt the Ephraimites, "Then they ſaid unto him, Say now Shibboleth; and he ſaid Sibboleth, for he could not ſo pronounce."

IBID. At Kiew, where Catharine II. remained three months, ſhe was received under triumphal arches; and, having heard the petitions of the deputies from diſtant nations, and extended the walls of the city, ſhe inſcribed, with an arrogant anticipation, the following motto in Greek characters on the quarter next to Conſtantinople; "Through this gate lies the road to Byzantium."

PAGE 76. From the books of the Cuſtom-Houſe it appears, that the trade of Aſtrachan, from 1793 to 1797, amounted to

744,100

744,100 roubles; eleven veſſels entered the port, and twenty-four cleared out. The collective trade of the Caſpian may be reckoned at more than three millions of roubles, the balance of which, againſt Ruſſia, is eſtimated at 1,850,000 r. *Storch.*

Ibid. We have been frequently ſtruck with a curious coincidence, (ſays Mr. Campbell,) of which the ingenious Mr. Bailli would not have failed to avail himſelf, had the fact reached his knowledge. That great philoſopher, it is well known, ſought for the celebrated *Atlantic* iſland in the northern hemiſphere. Now *Atl* is the Aſiatic name for the Volga, and *Atlanta* ſignifies in the Sanſcrit, the extremity of the Volga. The Volga and the Bazan form an iſland (called Dolgoi Oſtrov) on the Caſpian, in which ſtands Aſtrachan. *Atl* in the above language means *bottomleſs*. Whether this name be applied to the river from its great depth, we leave others to determine. The capital of Ebu Hankal's kingdom of Atl is manifeſtly Aſtrachan. *Aſiat. Reg.*

Page 107. It is curious to trace the attachments and averſions of mankind in regard to animals. Our author informs us, that the Gumrah reckon a dog an unclean animal. By the Jews they were conſidered in a degrading light, and were not ſuffered to come within the precincts of the temple. "Am I a dog," ſays the Philiſtine to David. "What! is thy ſervant a dog?" ſays Hazael, &c. Herodotus relates, that the females were buried in conſecrated cheſts. When Tritantæchmes was appointed to the principality of Babylon, he kept ſuch an immenſe number of Indian dogs, that four great towns were exempted from every other tax, but that of maintaining them. The Turks, we are told, extend their charity ſo far, as to bequeath legacies to dogs. In the Aſiatic Regiſter, we learn that the Perſees breed great numbers of dogs, and feed them regularly twice a day. Wherever they ſee a dog, they preſently call him and offer him food. The dogs on the iſland of Bombay, a few years ago, were many of them mad; whereupon an order was given by the governor for

killing

killing them all without exception. This order being known, the Perſees were greatly alarmed: they met together, and entered into a ſolemn league in defence of their dogs, and threatened to protect *their* lives at the riſk of their own.

PAGE 112. The following words, ſelected from Guelden-ſtædt's Vocabulary of the Ghazi-Kumuk and Akuſchae languages, may be entertaining to the linguiſt:

	Ghazi-Kumuk.	*Akuschac.*
God	Saal, B'ſaal	Zalla
Maſter	Uſaldu	Talkan
Father	Pu	Tutteſch
Mother	Ninu	Neſch
Brother	Uſu	Utzy
Son	Ers	Durha
Wife	Gerſa	Chunul adim
Daughter	Duſch	Durſſi
Man	Tſchu	Adim
People	Umat	Adimt
One	Zava	Za, Zaual
Two	Kiva	Quial, Kial
Three	Schamba	Abal, Hæval
Four	Mukva	Ohval
Five	Cheva	Chujal
Six	Rechva	Wreekal
Seven	Erulva	Uruzal
Eight	Maiva	Maizal
Nine	Urutſchva	Urbtſchanal
Ten	Atzba	Jæzal
Twenty	Kua	Hvajal
Thirty	Suiva	Abzale
Forty	Mukzalva	Auzale
Fifty	Chuzalva	Chuzale
Sixty	Rechzalva	Urekzale
Seventy	Erulzalva	Verzale
Eighty	Maizalva	Giachzale
Ninety	Urzalva	Urtſchemzale
Hundred	Durſchva	Darſchal.

NOTES.

Page 119. Gmelin commenced his travels in June 1768. In 1770 and 1771, he visited the different harbours of the Caspian, and examined with peculiar attention, those parts of the Persian provinces which border upon that sea. Actuated by a zeal for extending his observations, he attempted to pass through the western provinces of Persia, which are in a perpetual state of warfare, and infested by numerous banditti. Upon this expedition he quitted, in April, 1772, Ghilan, and owing to many difficulties and dangers, did not reach Sallian till December, 1773. Thence he proceeded to Baku and Kuba, where he was friendly received by Feth Ali. After he had been joined by twenty Uralian Kosaks, and when he was only four days journey from Kislar, he and his companion were arrested, on the 5th of Feb. 1774, by order of a petty prince, through whose territory he was obliged to pass. It was urged as a pretence for this arrest, that, thirty years ago, several families had escaped from his dominions, and found an asylum in Russia; adding, that Gmelin should not be released until those families were restored. The professor was removed from prison to prison, and at length, wearied out with constant persecutions, he expired on the 27th of July, at Achmed-Kent, at the early age of 30. It is said, that the prince demanded 30,000 roubles for his ransom. *Encyclopædia.*

Page 128. Struys says, the natives suppose Derbend to have been built by Alexander the Great, and thence to have been called *Scaher Iynan*. But there is no reason to think that Alexander ever was in that part of the country; much less that he built a city there. *Scaher*, or *Caher Iynan*, relate to a history far prior to that prince. I have in many places taken notice of a person named *Ion*, *Ionicus*, and *Iyna-Hellen*, who is supposed to have been the author of the Sabian worship, and from whom the ancient Ionians descended. *Caher Iynan* was a city built certainly by some of that family, and named after their common ancestor Zoroaster, who is styled *Iyna-Hellen*. The Sabian worship commenced about the time that the tower of Babel was erected.

Hellen,

Hellen, ſays Euſebius, was the ſon of a perſon who eſcaped the Flood, the ſame as Ham. *Bryant.*

PAGE 130. The oriental hiſtorians have mingled the little they know concerning the tranſactions of European nations, particularly concerning Alexander the Great, and his conqueſt of Perſia, with ſo many fabulous and incredible circumſtances, that hardly any attention is due to them. Though they miſrepreſented every event in his life, they entertained a high idea of his great power, diſtinguiſhing him by the appellation of Iſkender Sulkarnein, (*i. e.* the two-horned,) in alluſion to the extent of his dominions, which, according to them, reached from the weſtern to the eaſtern extremity of the earth. *Robertson.*

PAGE 134. Various names have been given to the Caſpian ſea. The Perſians called it Gurſen, from the old capital Gurgan; by the Greeks it was called the Hyrcanian; by the Tatars it is denominated Akdinghis, or the White Sea; by the Georgians it is termed the Kurtſchenſkian Sea; and in the Zend-Aveſta it was ſtyled Tſchekæt-Dæeti, Kam-roud, and Odjeſtan. It is ſaid to receive eighty-four large rivers beſides the Volga; and thither do all the floods of the Perſian and Tatarian rivers haſten, as it were, to their common centre. In the Caſpian there are no tides; but hard gales of wind, either from the N. or S. raiſe the water three or four feet, ſometimes more; and when theſe gales ceaſe, the water returns to its ordinary depth, with a prodigious current and turbulent ſea. The Ruſſians loſe many veſſels on the ſhoals of the north coaſt, for want of attention and ſkill. At Derbend there was formerly a landing-place for goods, but it is now under water. This is the worſt port, if it may be ſo called, in the whole Caſpian Sea, not only with regard to the loading and unloading of boats on the ſhore, but from the ſtony, foul ground in the road. From May to September, there are ſeldom any hard gales of wind, and therefore ſhips in Derbend road bringing the lower end of the ſouth wall to bear due W. may lie in ſeven fathoms water, which is about three-

three-fourths of an Englifh mile from the fhore. *Journal of a Russian Officer.*

PAGE 135. Capt. Woodroofe fays, the ftones of the walls of Derbend are, for the moft part, of two or three tons in meafurement, and fome yet more weighty. They are well jointed, regularly laid, and compofed of fea-fhells, (none of which appear on any of the Cafpian fhores,) very ftrongly cemented. When Peter the Great returned from his Perfian war, the greateft motive to triumph feemed to be the receipt of the filver key of Derbend which was delivered to him, and which, in a pompous manner, was carried before him when he entered Mofcow.

PAGE 139. The government of the Arabian empire had hitherto centred in one branch, namely, in the family of Muhammed. Soon after his death, the inheritance was made over to Ali; hence arofe the feparation between the Alites and Ommijades, which has continued for ten centuries, in all the bitternefs of religious hatred. Herder's Ideas, Part 4. *Schrœder.*

PAGE 145. *Tumann* is a term in the language of the Yufbec Tatars fignifying 10,000, and anfwering to the myriad of the Greeks. The Tatars count the troops by tumanns, as we do by regiments; and denote the grandeur of a prince by the number of tumann under his command. *Yus-bec* means 100 lords, owing to the number of principalities the country is divided into. *Chardin.*

PAGE 158. Feth Ali died on the 26th of April, 1789, and was fucceeded by his fon Achmed; who died on the 20th of November, 1790, by exceffive drinking. His brother Scheik Ali fucceeded.—*Schrœder.* Gmelin defcribes Feth Ali as very much given to drinking. He had fix wives, and was tolerably beloved by the inhabitants of Derbend. He levied duties according to circumftances: they were paid in kind, and confifted of horfes, cattle, barley, wheat, rice, and fruits.

His

His military force confifted of 40,000 men, compofed of Perfians and fubfidized Tatars.

PAGE 163. Sir W. Jones is of opinion, that one great fpring and fountain of all idolatry, in the four quarters of the globe, was the veneration paid by men to the vaft body of fire, which "looks from his fole dominion like the God of this world;" and another, the immoderate refpect fhown to the memory of powerful or virtuous anceftors, efpecially the founders of kingdoms, legiflators, and warriors, of whom the *sun* and the *moon* were wildly fuppofed to be the parents. The Perfians in India, as is generally known, pay divine adoration to *fire*, but not in a fenfelefs and idolatrous manner. For, we are affured that they worfhip it only as an emblem of the Divinity, and as his chief agent in the fyftem of the univerfe. They never extinguifh fire. They will ftand for hours by their lamps, putting up their prayers to God, with folded hands, and their eyes turned towards heaven with great marks of devotion. Zoroafter (fee the Zend-Avefta) confidering fire as the pureft fymbol of the ever operating Deity, it was natural that he fhould recommend its particular worfhip; and as fire is the only element which is not fenfible, except it be lighted, he commanded altars to be raifed, on which it was to be preferved. Thus fire became the moft common and moft impofing object of worfhip amongft the Perfians. We find Cyrus facrificing firft to Vefta, (fire,) and afterwards to Jupiter.

PAGE 164. Baku may be denominated the beft, if not the only true port in the Cafpian. In the mole are two fathoms water. The river Kura has at its mouth two fathoms water, and is a general rendezvous for all Perfian veffels going to, or returning from Baku. On the eaftern fide of the bay, they dig falt and brimftone, and coafting veffels receive their cargoes. *Journal of a Russian Officer*.

PAGE 169. Near Rainanghong in the Burmah dominions, (fays Capt. Cox,) are wells of petroleum. They are fcattered irregularly

irregularly about the downs, at no great diſtance from each other. At this particular place, we were informed, there are 180 wells, and four or five miles to the N. E 340 more. The oil is drawn pure from the wells, in the liquid ſtate as uſed, without variation; but in the cold ſeaſon it congeals in the open air, and always loſes ſomething of its fluidity; the temperature of the wells preſerving it in a liquid ſtate fit to be drawn. A man, who was lowered into a well 100 cubits, in my preſence, and immediately drawn up, perſpired copiouſly at every pore. The average produce of each well *per diem* may be eſtimated at 798 hhds.; and as there are 520 wells regiſtered by government, the groſs amount produce of the whole *per. ann.* will be 412,360 hhds.; worth at the wells 289,737 ſicca rupees. The oil is of a ding green, and odorous. It is uſed for lamps, and boiled with a little *dammer*, (a reſin of the country,) for paying the timbers of houſes, and the bottoms of boats, &c. Its medicinal properties known to the natives are, as a lotion in cutaneous eruptions, and as an embrocation in bruiſes and rheumatic affections. *Asiatic Register*, 1800.

PAGE 179. The probationary ſtate of the Saniaſſi in India far exceeds, in exquiſite tortures, that of the Gebers. The word *Saniassi* ſignifies the forſaking of all actions which are deſirable. By the Saniaſſi is properly to be underſtood the Brahmin in his fourth and higheſt degree of ſpiritual diſcipline. The leading principle that ſways the mind of the Saniaſſi is by unexampled auſterities to ſubdue the body, becauſe he is convinced that ſubjugation of the paſſions will neceſſarily follow. On entering this degree, he inſtantly, and without ſcruple, diſcards for ever the deareſt friend, and the tendereſt relative. He throws away every article of dreſs, except a ſcanty linen cloth of a yellow colour, which girds his waiſt; and, with a pitcher in one hand and a pilgrim's ſtaff in the other, he hurries away to the deſert never to return. Famine and miſery are the companions of his ſolitude. Abſorbed in profound meditation on the Deity, he never violates the ſacred ſilence in which his lips are ſealed, except to pronounce

pronounce the myſtic word *Awan*, which is the commencement of the Vedas. His food is the fruits and herbage that ſpontaneouſly ſpring up in the deſert: if theſe fail, the laws of his ſevere order permit him to go to the neareſt village, and beg a handful of boiled rice, or other food, which he eats on the ſpot; if they throw it on the ground, he takes it up with his mouth, ſwallowing only as much as will ſerve to ſuſtain life. The rains which, during the annual inundations, deſcend in tropical regions with ſuch reſiſtleſs violence, and ſweep every thing before them, moleſt not the inflexible devotee; nor is the naked northern Anchorite obſerved to ſhiver amidſt the inceſſant ſnows that fall upon the ſummits of Heemacote, and encircle, up to his neck, the human ſtatue in the holy mountains of the Brahmins. In fine, the higheſt diſtinction, to which the Saniaſſi aſpires, is a ſtate of invincible apathy. By long habits of indifference, he becomes inanimate as a piece of wood or ſtone; and though he mechanically reſpires the vital air, he is, to all the purpoſes of active life, *defunct*. In conſequence of theſe unexampled ſeverities, and this invincible abſtraction from every thing finite, the veneration which the whole Indian nation entertain for the Saniaſſis is beyond all conception. Viſhnu himſelf reveres them; to whatever object they touch, they impart ſanctity, and the very duſt of their feet is conſecrated, from the ſteeps of Caucaſus to the point of Comorin!!! *Maurice Ind. Ant.*

PAGE 180. The violence of the S. E. wind is extremely dangerous on the N. and N. W. coaſt of the Caſpian, as it cauſes great inundations. On the 21ſt of October, 1790, the water roſe ſo high, that Aſtrachan was nearly inundated, and the ſurrounding deſert was overflowed to the extent of 120 verſts. *Reineggs.*

PAGE 183. There is very ſolid ground for believing, that the Ahwans, or Afgans, deſcended from the Jews; becauſe, they ſometimes in confidence avow that unpopular origin, which, in general, they ſedulouſly conceal, and which other Muſſelmans positively assert; and, principally, because their language is evidently a dialect of the Scripture Chaldaic. *Jones.*

SPECIMENS of the AHWANIC LANGUAGE.

God	Chudai	Six	Spay	Twenty	Schil
Master	Agoi, Malik	Seven	Ue	Thirty	Desch
One	Jau	Eight	Ate	Forty	Salvecht
Two	Dua	Nine	Nehe	Fifty	Pansos
Three	Dre	Ten	Las	Hundred	Sil
Four	Salur	Eleven	Jau las	Thousand	Sir.
Five	Pinse	Twelve	Du las		

GUELDENSTÆDT.

IBID. Cambysena, according to Strabo, was a country of Albania, between the rivers Cambyses and Cyrus, and the mountains. It was one of the Prefectures of Armenia.

PAGE 200. This Persian aphorism evidently implies, that however potent any army may be, the situation of the mountains of Daghestan is such, and the people are so brave, that the success of an expedition against them must be very precarious. Nadir Shah, whom no Persian king exceeded in vanity and ambition, determined to try his fortune against them. At the end of the year 1741, he marched, at the head of 35,000 men, into Schirvan; and from thence directed his course into the mountains of the Lesghaes. In order to facilitate his progress, he caused vast quantities of wood to be hewn down on both sides of the road, designing to frustrate any attempt which the Lesghaes might make to lie in ambuscade; for, he was not ignorant that they are as crafty as they are brave. These people defended themselves against the conqueror of India, and, after repeated defeats, Nadir saw, though too late, that he committed a false step. *Hanway.*

PAGE 215. Amongst all the nations under the whole canopy of heaven, there are none so barbarous and cruel, none so utterly lost to all the sentiments of humanity and civility, but that have embraced and continued amongst them the notion of a Deity, or ſome being that ought to be adored by them. The numerous idolatries of the Egyptians, Greeks, and Romans, are too well known. In the iſland of Java, they worſhip whatſoever it is that they firſt meet, or chance to ſee in the morning; and pray unto it all the day, although it be a hog, or any other animal or reptile. In the province of Manta, in Peru, they worſhip an emerald. The Alani, in S. America, amongſt other barbarous ceremonies, fix a naked ſword in the ground; and this they religiouſly worſhip as the god of thoſe regions, where they make their abode. The old Iriſh at every change of the moon, worſhipped her, bowed, and made their ſupplications, and with a loud voice, thus ſpake unto the planet: "We pray thee leave us in as good a ſtate as then haſt found us." *Johnston's Gen. Hist. of Man.*

Page 216. Counſellor Gatterer is of opinion, that the Leſghaes are of Finnic origin, as they ſpeak ſix different dialects of that language. Theophanes knew them by the name of Legæ; he ſays, that they bordered on Albania, and as they inhabit the ſame country to this day, they may very well be proud of their anceſtry. *Schrœder.*

Page 222. By the ſame method, nearly, two Kirgiſes often carry off the horſes of a whole Ruſſian regiment. Namely, they ſneak up cloſe to the place, in the deſerts, where the horſes are feeding, and, though under a ſtrong guard, ſpring with the greateſt agility through the piquets, diſcharge their piſtols on all ſides, and gallop off. The horſes become terrified, and where one runs, the reſt follow. *Schrœder.*

Page 224. The German Editor, by his note of admiration, appears much aſtoniſhed that the Leſghaes ſhould eat human fleſh. But *Androphagi* are not uncommon. Herodotus ſpeaks of men-eaters, a ſeparate nation, and by no means Scythian. Galen ſays, he was informed by ſuch as had made trial thereof, that the fleſh of man is in taſte much like unto that of beef. Thevenot ſays of Debea, in the province

vince of Guzerat: "The inhabitants of this town were formerly ſuch as are called *Merdi-Coura*, men-eaters; and it is not many years ſince human fleſh was there publicly ſold in the market." In the deſcription of *Montanna Real*, in S. America, we read of the wild Caſibos and Guaguas. The greateſt ornament of the latter is a necklace made of the hearts of their murdered enemies. Nothing is ſo ſavoury to their taſte as human fleſh. They ſalt and dry it in the ſun. The Capanagnas eat their dead, and think they cannot honour them more highly. Artemiſia, it is ſaid, mingled the aſhes of her huſband with her drink. Polyphemus is thus deſcribed in Virgil:

"The entrails of ſlain wretches are his food,
"And then he quaffs their black and lukewarm blood."

De Pons (in his Travels through S. America) ſays, that the Spaniards have been ſo convinced that the only method of imbibing the Indians with a love for their religion would be to chriſtianize their taſtes and habits. The queſtion, therefore, has been agitated, "whether it were permitted to eat human fleſh; and what adds to the ſingularity of the queſtion is, that it has been decided in favour of the *Androphagi*. The biſhop of Montenegro gives his opinion, "that in caſes of neceſſity, the Indians may eat human fleſh without any ſin attaching, becauſe it is not an evil in itſelf." This ſtrange deciſion has been made in, perhaps, the moſt fertile part of the globe, covered with foreſts which furniſh an inexhauſtible ſupply of game, and watered by rivers full of fiſh, turtle, &c. The Mexicans ſacrificed their priſoners of war in the temple of the god of battles. The prieſts, it is ſaid, afterwards ate them, and ſent portions to the emperor, and the principal lords of the realm. When peace had laſted ſome time, the prieſts took care to have it inſinuated to the emperor, that the gods were hungry; and war was commenced with no other view than to make priſoners. *Raynal.*

PAGE 229. According to De Guignes, the Awars, who acted ſo conſpicuous a part in the hiſtory of the middle ages,

 and

and who ſettled in Dacia and aſſiſted in the deſtruction of the kingdom of the Gepidæ, 567 years P. C. might be of Caucaſian origin. One of their moſt powerful leaders was Bajan Khan, a ſecond Attila.—*Schrœder*. Peyſonnel ſays, that when the Awars firſt entered Conſtantinople, the inhabitants went out to examine them, never having ſeen ſuch ſtrange beings before.

Page 233. With regard to the Ezechs, perhaps the caſe is reverſed, ſays the German Editor. A tribe of the Ezechi ſettled in Bohemia. At firſt they were under the dominion of the Franks, and after the year 596, under that of the Awars. See Gelaſii epiſt. quâ gentis Czechicæ origo, &c. Pragæ, 1767 *Schrœder*.

Page 238. The Oſſi appear to me, ſays Gueldenſtædt, to be a remnant of the ancient Polovs or Uzes, which their language and common names tend to confirm, and is not improbable from hiſtory. In 1110, when the Polovs were beaten by the Ruſſians on the Don, a remnant, the preſent Oſſi, fled to the mountains. The Tatars call them *Kuscha* (bones.) They are ignorant of religion and laws, rude, rather plunderers than thieves, falſe, crafty, and faithleſs to one another.

Page 247. The Japoneſe appear to carry their ſpirit of revenge to a great length. Titſingh, in his hiſtory of Japan, ſays, that no native is inſulted without the blood of his aggreſſor expiating the injury; or, in caſe that he cannot have ſatisfaction, he kills himſelf. This ſpirit of vengeance extends even to females. They, as well as the men, conſtantly carry a dagger in their girdles, and employ it with the utmoſt coolneſs in their perpetual quarrels, not only againſt their enemies, but their huſbands and brothers, and even againſt themſelves.

Page 255. Owing to the inundation of the Terek in 1785, when the road along its banks was deſtroyed, the three forts were abandoned. Hence the Oſſi, and other mountaineers, became

became uncontrolled, and will now purfue their favage manner of living for a long period unmolefted. *Reineggs*.

PAGE 260. Captain Wilford informs us, that the city of Bamiyan, fituated on the road between Balkh (the ancient Bactria) and Cabul, is remarkable for a vaft number of curious receffes, cut out of the folid rock; fome of which, from their dimenfions, are fuppofed to have been temples. Coloffal ftatues, one of 240 feet high, the other 150, both perfectly erect, and fixed to the vaft rock out of which they are cut, attract and furprife the traveller.

PAGE 266. It feems agreed, (fays Sir W. Jeans) that the fingular people, called Egyptians, and, by corruption, gipfies, paffed the Mediterranean immediately from Egypt; and their motley language, of which Mr. Grellman exhibits a copious vocabulary, contains fo many Sanfcrit words, that their Indian origin can hardly be doubted. The authenticity of that vocabulary feems eftablifhed by a multitude of gipfey words, for which the collector of them could find no parallel in the vulgar dialect of Hinduftan, though we know them to be pure Sanfcrit, fcarce changed in a fingle letter A very ingenious friend, to whom this remarkable fact was imparted, fuggefted to me, that thofe very few words might have been taken from old Egyptian, and that the gipfies were Troglodytes from the rocks near Thebes, where a race of banditti ftill refemble them in their habits and features. But, as we have no other evidence of fo ftrong an affinity between the popular dialects of old Egypt and India, it feems more probable, that the gipfies, whom the Italians call *Zingari*, (and the Germans *Zigeuner*,) were no other than Zinganians, as Mr. D'Anville alfo writes the word, who might, in fome piratical expedition, have landed on the coaft of Arabia or Africa, whence they might have rambled to Egypt, and at length have migrated, or been driven to Europe.

PAGE 274. According to Peyfonnel, the Tfcherkaffian, a very ancient language, which is ftill ufed throughout Circassia and the Kabardah, from the sea of Asov to the Caspian, resembles no other language in the world, except the Hungarian, with which it has a certain affinity. It may be, that it was formerly common to the Jazyges, who inhabited the northern coast of the Palus Mæotis, and that the Jazyges Metanastes, the first Scythians who invaded Pannonia, carried it there. It may be, likewise, that it was introduced by the Hungarian Turks who came after them, and who were mixed with the Chazares, and many other people of Circassia. In fact, the skeleton (if I may use the expression) of the modern Hungarian, has not the least connexion with any known language, except the Circassian.

Page 291. The following extracts from Mr. Brunton's Letters will give the reader an idea of the present state of Kabardah and of the Tscherkassians. He represents the latter as a fierce and cruel people, who never scruple to kill and plunder any whom they happen to meet, connected with the Russians, if they are able to master them. It seems that the Kabardins and Russians have never yet come to an agreement. The Emperor of Russia has granted the Missionaries 6000 *dessatines* (somewhat more than 1600 acres) of land, which have been measured off to them. The terms of the grant are much more favourable than have ever been given to any other colonists. The natives feel their inferiority to the Missionaries in the useful arts of life, and look up to them with respect; their violent prejudices against Christianity have considerably abated; and even some of the Effendis have become friendly to the Missionaries. The plague has raged around them for two or three years past. Agues are sometimes very severe. Cancers and epilepsies are pretty frequent among the Mahommedans. Great numbers of them are affected with madness, which they suppose to proceed from the influence of the devil. The priests whip them severely; threaten them till they tell the names of those whom they suppose to torment them, which are written and thrown into the fire; and read over them a small Arabic book containing the names of Noah, Mohammed, Jesus Christ, &c. by which means

means they conceive themſelves to reſtore many to ſoundneſs of mind.

The Kabardins pretend to have come from Egypt to Conſtantinople, from Conſtantinople to the Crimea, and from the Crimea to their preſent country. The tyranny of the Moſlems, they ſay, was the cauſe of their peregrinations. They are almoſt all thieves and robbers. Moſt of their chiefs are robbers, and employ the moſt fearleſs of the people under them to commit depredations, and carry away the property of their neighbours. The number of Ruſſians, whom they murder from time to time, might ſoon form a people more numerous than themſelves.

PAGE 293. Rubruquis, who was ſent by St. Beidis, in 1253, to the Great Mogul, and Haithon of Armenia, who wrote his Armenian hiſtory about the year 1307, ſay, that the country called *Cumania* extended from Caucaſus to the Volga, and contained the ruins of the city of Saray. *Schræder.*

PAGE 295. The Bejs and Mirſas have authority over all who are inferior to them in rank. The Kabardin Bejs are rich; the property of ſome of them in ſheep, mares, and ſlaves, amounts to 120,000 or 130,000 roubles. Many of the Tatar Mirſas, however, are ſcarcely able to buy wives for themſelves; but the people pay particular reſpect to them. Some of the Kabardah free people are rich, yet they are always dependant upon the Bejs. From this dependance, no degree of wealth can procure them exemption. *Brunton.*

PAGE 296. It may appear ſomewhat ſurpriſing, ſays Dr. Grieve, that one of the moſt important productions of milk ſhould remain unknown to the moſt enlightened parts of Europe. The production I mean is, the vinous liquor which is procured by fermentation from mare's milk (*kuemiſs*.) The following method of making it is common among the Baſchkir Tatars. Take of freſh mare's milk, of one day, any quantity; add to it a ſixth part of water, and pour the mixture into a wooden veſſel. Uſe then, as a ferment, an eighth part of the foureſt cow's milk that can be got; but, at any future preparation, a ſmall portion of the old kuemiſs will better anſwer the purpoſe of ſouring. Cover the veſſel with a thick cloth, and ſet it in a place of moderate warmth; leave it at reſt twenty-four hours, at the end of which time the milk will have become ſour, and a thick ſubſtance will be gathered at the top; then, with a ſtick, made at the lower end in the manner of a churn-ſtaff, beat it, till the thick ſubſtance above-mentioned be blended intimately with the ſubjacent fluid: in this ſituation leave it again at reſt for twenty-four hours more; after which, pour it into a higher and narrower veſſel, reſembling a churn, where the agitation muſt be repeated as before, till the liquor appears to be perfectly homogeneon and in this ſtate it is called *kuemiss*; of which the taſte ought to be a pleaſant mixture of ſweet and ſour. It has been tried, as a medicine, with great ſucceſs, in nervous diſorders, incipient phthiſis, abſceſs, &c.

PAGE 299. On account of the diſturbances in Caucaſus, Iſlam Murſa retired in 1789, further N. to the country of Beſch Tow,. about eighty verſts from the river Kuban, and placed himſelf under the protection of Ruſſia. *Reineggs.*

PAGE 303. The iſland of Taman is at preſent re-peopled, owing to the late Prince Potemkin, by Koſaks from the Black Sea. In April, 1793, a tribe of them, conſiſting of 17,500 males, took poſſeſſion of the iſland, and received ploughs and other inſtruments of huſbandry for cultivating the land. See the Ruſſian Journal. *German Editor.*

PAGE 304. The author is miſtaken in placing the city of Phanagoria on the right bank of the Achardeus. All hiſtorians agree, that it was in the iſland of Taman, and it probably ſtood in the neighbourhood of the modern city of the ſame name. This is confirmed by D'Anville's map, *Orbis Romani pars orientalis*, and by the map of the Black Sea in the Travels of Anacharſis. *German Editor.*

PAGE

NOTES.

PAGE 317. I am not certain, ſays Profeſſor Pallas, what nation has given to this mountain the name of Ellbrus, or Ellborus, which perhaps is not inferior to Mount Blanc. The Tſcherkaſſians call it *Oshha makhuer*, or the Happy Mount; and the Akeſes *Orſi Ipgub*. They conſider it as the reſidence of the king of the hobgoblins, whom they call *Dschin-Padishah*. In the viciſſitudes of an unfortunate war, this mountain is an aſylum to thoſe nations.

PAGE 323. The port of Gelintſchik Limoni, which receives two rivulets, is ſhut in by the mountains. Its mouth is about a thouſand toiſes in width, and is defended by a battery of four or five ſmall guns, with a garriſon of twenty-five men. The country is in general very mountainous, interſected by ravines. We find in the woods patches of ground in cultivation, and huts ſcattered about in the valleys. The land is generally fertile, ſupported by beds of rocks, mixed with argill, in a vertical direction. *Le Chevalier's Travels.*

PAGE 326. *Eristaw* is an old Georgian word ſignifying "head of the people," from *Er*, people who inhabit the mountains, and *Tawi*, head, or governor of a diſtrict of mountaineers. As this power was formerly entruſted to perſons of rank, who afterwards abuſed it, and threw off their allegiance; the preſent Zaars of Georgia and Iberia have been obliged to divide the power of an *Eristawi* amongſt different adminiſtrators. *Reineggs.*

PAGE 328. The Abghazs, or Abghazſians, always go armed with a ſabre, gun, and piſtols, or arrows, on account of their princes being conſtantly in a ſtate of warfare. Their dreſs conſiſts of a pair of wide pantaloons, a ſmall waiſtcoat with ſleeves, and a kind of long robe faſtened by a leathern girdle. They let their beards grow, but ſhave the head; and wear a kind of kalpak. In war time, they wear a coat of mail, which reaches to the knees.

From certain ceremonies, ſuch as placing croſſes on the graves, carrying food to the dead, fixing hooked ſtakes in the church

church-yards, that the dead may hang their horfes to them when they return from their nocturnal peregrinations, &c. one might fuppofe that their religion was a ftrange mixture of Chriftianity, Mahommedanifm, and Paganifm. Some of their laws and cuftoms are invariably followed; as for example, a flave who had been ftolen, muft be reftored wherever he is found, and the thief muft defray all the expences of the fearch. Thofe whom they take under their protection are in perfect fecurity. Hence the Turkifh troops are always accompanied by one or two Abghazs when they march through the country. One of them, engaged to accompany the French engineer Lafitte, being prevented, fent him his *arrow*, ftating that it would infure his protection whenever he fhowed it. Their trade confifts of flaves of both fexes, wax, honey, butter, fkins, wool, and furs; which they barter for falt, linens, coarfe ftuffs, guns, fabres, piftols, and particularly powder. *Le Chevalier.*

PAGE 330. It is remarkable, fays Dr. Horne, that, in the earlieft ages, a cuftom fhould be found to prevail, both among the people of God, and idolaters, of fetting apart and confecrating gardens and groves, for the purpofe of religious worfhip. Thus Abraham planted a grove at Beerfheba. The worfhippers of falfe gods are defcribed, in the writings of the prophets, as "facrificing in gardens." A furprifing uniformity in this point may be traced through all the different periods of idolatry, as fubfifting among the Canaanites, Egyptians, Greeks, and Romans. Groves were dedicated to the gods, and particular fpecies of trees were facred to particular deities. The fame ufage prevailed among the Druids, in thefe parts of the world. And to this day, the aifles of our Gothic churches and cathedrals are evidently built in imitation of thofe arched groves, which of old fupplied the place of temples.

PAGE 331. The gloomy cavern and the confecrated grove bore witnefs to the earlieft devotions of mankind. From the deep obfcurity of caverns and forefts have, in every age, iffued the

the light of philoſophy, and the beams of religion. Zoroaſter, amidſt the gloom of a cavern, compoſed his celebrated ſyſtem of theological inſtitutions. The renowned philoſophers, Epictetus and Pythagoras, ſought wiſdom in the ſolitary cell. Even the venerable prophets of the true religion took up their abode in the ſolitudes of the deſert; and the herald of the Meſſiah declares himſelf to be "the voice of one crying in the wilderneſs, &c." In later ages, the crafty Mahommed thought it neceſſary to ſhun the ſociety of men, and retired to fabricate his daring impoſtures in a lonely cave. *Maurice.*

Page 332. Chardin gives a curious account of ſacrificing camels. In Iſpahan, on the firſt day of the month, a camel is taken from the royal ſtables, and led in proceſſion, accompanied with muſic. It is dreſſed up with flowers. Some offer it graſs; others pluck off the hairs, which they keep as relics. It is led into the houſes of the grandees, that the women may ſee it. Without the city is a large ſquare, called "the place of ſacrifice." When the king arrives, the camel is made to lie down with its head turned towards Mecca, and its head and legs are tied together. The prieſts, on horſeback, are arranged on one ſide; the king, with his nobles, on the other. When the former have finiſhed their prayers, the grand prefect advances on horſeback very near to the animal. He is then preſented with a ſhort ſpear, well poliſhed, and ornamented with gold, and ſtrikes it into the left ſhoulder, invoking at the ſame time, in an audible voice, bleſſings on the king and nation. The camel is then cut into ſix pieces, according to the diviſions of the city.

Page 333. At the time of Herodotus, the Lazi lived on the lake Mæotis. Thence they appear to have retired to Pontus, and ſettled in the country afterwards called Zighia; till they proceeded to the Phaſis, and drove out the Colchians. Pliny is the firſt who mentions that the Lazi lived on the Phaſis. Since his time, they inhabit the ſame country, as is clearly proved by the Byzantine hiſtorians. Colonel Gærber is of opinion, that the Lazi and Leſghaes are the ſame people, but he is miſtaken. *Schrœder.*

Ibid.

Ibid. The Alani were of Aſiatic origin, but repreſented by A. Marcellinus as more poliſhed in their cuſtoms and manners. Joſephus deſcribes them as Scythians, who dwelt between the Tanais and the lake Mæotis. From Mr. de Guignes' Hiſt. Huns, who has taken pains in inveſtigating their origin and hiſtory, we learn that the name *Alin* ſignifies mountain, and that theſe people derived their appellation from the mountains which they inhabited. According to Gibbon, after the Alani were defeated by the Huns, a colony of exiles found a ſecure refuge in the mountains of Caucaſus, where they ſtill preſerve their name and independence.

Page 335. By our author's deſcription, the Soani are as dirty a race as they were in Strabo's time. He deſcribes them (in B. xi.) as inferior to none in filthineſs, but far ſuperior in power and courage. They have, ſays he, a king, and a council of three hundred men; and it is ſtated, that they can bring 200,000 men into the field, for they are a nation of warriors. They have an admirable method of poiſoning the ends of their ſpears.

Page 336. Certain tribes of religions in India, (like the Soani,) are chiefly diſtinguiſhed in their appearance by the exceſſive length and impenetrable entanglement of their hair. As they reject the luxury of moiſtening it with oil, (which the natives in general uſe,) it becomes ſun-burnt, and of a brick-duſt hue. The whole forms a diſguſting ſpecies of turban; nor have they any other cover to ſhelter their heads from the intenſe heat of the ſun, and inclemency of the weather. *Hindoo Literature.*

Page 344. The Zaar Salomo died on the 23d of April, 1784. Dawud, his nephew, was choſen king, and in July, 1790, was dethroned by his youngeſt couſin. He returned to Iberia in 1791, abdicated the crown, and received as a maintenance his paternal poſſeſſions. *Reineggs.*

NOTES.

PAGE 347. The Mingrelians have many ſacrifices ſimilar to the Jewiſh; but, one in particular at the time of vintage. They fill a large jar, containing about twenty flaſks, with the beſt wine; this is ſet apart as an offering to St. George. When St. Peter's day comes, the maſter of the houſe carries a cup of it to St. George's church, and ſays his prayers. On his return, he and his family go into the cellar, lay a loaf with cheeſe, onions, or leeks, on the conſecrated jar, and pray together. They then kill a calf, a kid, or a pig, and the blood is poured around the jar by the maſter of the houſe. After praying again, they begin to eat and drink. *Chardin.*

PAGE 351. In proceſſions, or for meetings, and on parochial buſineſs, the Mingrelians found a trumpet after the manner of the Jews: "But in aſſembling the congregation, ye ſhall blow without an alarm. And the ſons of Aaron the prieſt ſhall blow the trumpet."—*Numb.* x. When any article of value has been ſtolen from a church, they blow very loud; that, as they ſay, the ſound may alarm the thief, as if it were the voice of God; and that his conſcience may reproach him: "For he heard the ſound of the trumpet, and would not be admoniſhed; therefore his blood ſhall be upon him: but he that receiveth warning, ſhall ſave his life." *Ezek.* xxxiii. *Chardin.*

PAGE 352. It is no error in Strabo's placing the ſource of the Phaſis in the Armenian mountains; for, at his time, and long afterwards, the kings of Armenia reigned over that part of Caucaſus.—*Schrœder.* Gibbon deſcribes the Phaſis as deſcending from the Iberian Caucaſus with ſuch oblique vehemence, that, in a ſhort ſpace, it is traverſed by 120 bridges. Nor does the ſtream become placid and navigable, till it reaches the town of Sarapana, five day's journey from the Cyrus. As it ſucceſſively collects the ſtreams of the plain of Colchos, the Phaſis moves, with diminiſhed ſpeed, though accumulated weight. It can be navigated for forty miles by large veſſels.

PAGE

NOTES.

PAGE 362. Great drinkers are to be met with in all ages and nations. Alexander the Great, who was this way ſufficiently addicted, after the burning of Calanus, propoſed a prize of drinking, wherein he that drank moſt, ſays Plutarch, was one Promachus, who took off four *congies*, or forty pints. It is ſaid of the Emperor Maximinus, that he often drank, in one day, an *amphora* of the capitol, which is nine gallons Engliſh. It was a kind of uſual rule amongſt the Romans to drink down the evening, and to drink up the morning ſtar. A few years ſince, ſays Lipſius, at the wedding of a nobleman in Bavaria, a drinking wager was propounded amongſt the ſervants and retainers, and thither came one who drank little leſs than ſix gallons in a ſhort ſpace, and ſo went away with the prize. *Johnston's Gen. Hist. of Man.*

PAGE 364. On the 20th of October, Dr. Reineggs delivered the order of Alexander Nevſky to the hereditary prince of Georgia. And on the 18th of Nov. that of St. Andrew to the Zaar Salomon, which the late Empreſs ſent him by our author. *Schrœder.*

PAGE 365. Ghomi, or gom, is about the ſize of coriander ſeed. It is ſown in ſpring like rice. A hole is made in the ground with the finger, a grain is put in, and then covered. It produces a reed, like a ſugar-cane, about the thickneſs of one's thumb, which grows nearly ſix feet high, with an ear containing above 300 grains. It is gathered in October, and laid on hurdles to dry. It is inſipid and heavy. In dreſſing it, when the water begins to boil, it is gently ſtirred with a wooden ſpoon. As the grains diſſolve, and form into a paſte, the fire is leſſened, ſo that the water boils away, and the paſte dries in the veſſel. It is ſometimes as white as ſnow, and is ſerved up with ſmall wooden ſpoons made on purpoſe. Its qualities are extremely cold and laxative; and when it has been ſuffered to cool, or is heated again, it is good for nothing. The Circaſſians, the Mingrelians, the Georgians tributary to Turkey, Abghazians, Caucaſians, and all thoſe who inhabit the coaſts of the Black Sea, from the ſtraits of the

Mæotic

Mæotic lake round to Trebiſond, have no other kind of bread. *Chardin.*

PAGE 371. Pityus was deſtroyed by the Goths in their firſt naval expedition, in the third century. The city was provided with a convenient port, and fortified with a ſtrong wall, and conſtituted the utmoſt limits of the Roman provinces. Arrian places Dioſcurias, or Sebaſtopolis, forty-four miles E. of Pityus. *Gibbon.*

PAGE 384. I am of opinion, ſays our author, that ſome of the Greek emperors who long reigned over Georgia, by means of their viceroys, were accuſtomed (as I mentioned with regard to the Iberian monarchs) to unite the governorſhip of a province with certain ſtate dignities, and that they once granted them to a reſpectable individual of their *cartulariorum.* Hence, probably, has ariſen the modern name Kartel.

PAGE 388. Major Rennel places Ghori, or Gori, in the country of the Alarodians, forming the third diviſion of the eighteenth Satrapy. But, he ſays, we cannot find any authority for it; yet may ſuppoſe their country to be parts of Iberia and Albania.

PAGE 389. Krzinwal is alſo written Krzilowa, and takes its name from Krzila, the aſh-tree, (*faxinus excelsior,*) which grows there in great abundance. *Reineggs.*

PAGE 392. Except a few detached idols, (ſays Sir W. Jones,) the only great monuments of Tatarian antiquity are a line of ramparts on the weſt and eaſt of the Caſpian, aſcribed indeed by ignorant Muſſelmans to Yajuj and Majuj, or Gog and Magog, that is to the Scythians; but manifeſtly raiſed by a very different nation, in order to ſtop their predatory inroads through the paſſes of Caucaſus. Gibbon ſays, that the imaginary rampart of Gog and Magog, which was ſeriouſly explored and believed by a caliph of the ninth century, appears to be derived from the gates of Mount Caucaſus

ſus and a vague report of the wall of China. But the real paſs of Jajuje and Majuje is fixed by Major Rennel (ſee his map) in the mountains of Imaus, the weſtern declivity of the high level of Aſia.

PAGE 402. On the Armenian language, (ſays Sir W. Jones,) which I never ſtudied, becauſe I could not hear of any original compoſitions in it, I can offer nothing deciſive; but am convinced, from the beſt information procurable in Bengal, that its baſis was ancient Perſian, of the ſame Indian ſtock with the *Zend*, and that it has been grȃdually changed ſince the time when Armenia ceaſed to be a province of Irân. The letters, in which it now appears, are allowed to be comparatively modern. Moſes Choronenſis, than whom no man was more able to elucidate the ſubject, has inſerted in his hiſtorical work, a diſquiſition on the language of Armenia, from which ſome curious information may be collected.

IBID. The famous Zaar Herakleus, of whom we have heard ſo much, was formerly a Perſian vaſſal, but had regained his independence. He defended it afterwards ſucceſsfully againſt the Turks, by the aſſiſtance of Ruſſia; yet it coſt him dear. In the Perſian war, he was obliged to acknowledge himſelf as the late Empreſs's vaſſal, and receive a Ruſſian garriſon in Teflis. Many petty neighbouring princes, at the ſame time, ſubmitted to the ſame fate, and figured in the Ruſſian calendar under the title of Zaars or Kings under the protection of the empire. *Maſson.*

PAGE 406. Mr. Tooke remarks, that the term *Kaſak* is generally taken in a bad ſenſe. Captain Kirkpatrick ſays, "I think I recognize the Koſaks, or Coſſacks, in the oriental term, or appellation, of Cozâck, or Cuzzâck, which is applied generally to any *banditti* or free-booters: but which I have alſo met with, in a reſtrained ſenſe, and applied to a particular tribe, or nation." According to Storch, the word is probably Tatarian, and means an armed warrior. He divides them as follows:

Little Russian Kosaks.

Slobodiſian	collateral branches.
Saporogiſhian	

Don Kosaks.

Volgaiſian, (Dubovſkiſian, Aſtrachanic.)	collateral branches.
Grebenſkiſian or Terkiſhian. (Semeiniſian.)	
Orenburgian.	
Uralian (formerly Jaikiſhian.)	
Siberian	

NOTES

TO VOL. II.

Page 3. Mount Ararat is ſtill called Aghridagh, Aghirdagh, (*Zagrus*, Strabo,) and Maſis, (*Masius*, Plin.) conſequently, the general name of a nation, the Getæ, was derived from the country they inhabited. Strabo mentions alſo the *Massabatæ*, which means nothing more in the preſent vernacular tongue, than "the inhabitants of Maſis or Ararat."

Reineggs.

Page 4. Moſes Choronenſis learnt the Greek language in Alexandria, and travelled through Paleſtine, Italy, and Greece. In his old age he compiled a hiſtory of Armenia, in three books; collected not only from other hiſtorians, but alſo from records in the archives, which he did not employ, however, with ſufficient judgment. He left, likewiſe, a geography of the then known world, which is merely an extract from the Chorography of Pappus of Alexandria. Yet it is become more valuable, owing to the original being loft. *Meusel.*

Page 18. A piece of fuperftition, of which no traces are to be found in any other country, exifted among the Mexicans. On certain days, the prieft made a ftatue of pafte, which was fent to be baked; they then placed it on an altar, where it became a divinity. Upon this day innumerable crowds flocked to the temple. The priefts cut the ftatue in pieces, and diftributed a portion to all the perfons in the affembly, who ate it, and thought they were fanctified by fwallowing their god.—*Raynal.* The reader will eafily trace, from the above, the *host* of the Roman church. In the twelfth century (fays Clarkfon) a notion prevailed, that confecration by a prieft had the power of abolifhing the fubftance of the bread, and of fubftituting the very body of Jefus Chrift. This was called the doctrine of tranfubftantiation. This doctrine appeared to Luther, at the dawn of reformation, to be abfurd; and he was of opinion, that the Sacrament confifted of the fubftance of the bread and wine, or in other words, that the fubftance of the bread remained, but the body of Chrift was inherent in it; fo that the fubftance of the bread, and of the body and blood of Chrift, was there alfo. This was called the doctrine of confubftantiation, in contradiftinction of the former.

Page 23. The Zend was the language of Georgia, of Irân Proper, and of Aderbedjan the country of Zoroafter, fome centuries before Chrift. The grammatical terminations, peculiar to the Zend and Georgian, prove an immediate refemblance between the two. The Parfi, or pure Perfian, is derived immediately from the Zend, which was in ufe under the firfts monarchs of Perfia. *Du Perron.*

The following words are from Gueldenftædt's vocabulary of the modern Georgian:

God

	Kartel.	Mingrelia.
God	Gmerti	Gorunti
I	Me	Ma
Thou	Schen	Si
He	Is	Itina
We	Tſchuen	Tſchſchini
Ye	Tkuen	Tkva
They	Iſſini	——
What?	Ra	Mugoko
Yes	Ho	Ko
No	Ara	Var
One	Erti	Arti
Two	Ori	Shiri
Three	Sami	Sumi
Four	Otſchi	Antſchi
Five	Chuti	Chuti
Six	Ekuſi	Amſchi
Seven	Schuedi	Skoiti
Eight	Rva	Rua
Nine	Zchra	Tſchchora
Ten	Ati	Viti
Twenty	Ootzi	Etſchi
Thirty	Oozd-ati	Etſch da viti
Forty	Ormoodzi	Dſcharnetſchi
Fifty	Ormoodzdati	Dſcharnetſchdaviti
Sixty	Samootzi	Sumenetſchi
Seventy	Samootzdati	Sumenetſchdaviti
Eighty	Otchmootzi	Autſchinetſchi
Ninety	Otchmootzdati	Atſchinetſchdaviti
Hundred	Aſſi	Oſchi.

PAGE 37. The government of Caucaſus is divided into two provinces; viz. the province of Caucaſus, and the province of Aſtrachan. The former conſiſts of the following diſtricts:

Ecatherinograd, the capital of the government, on the river Malka; diſtant from Peterſburg 2528 verſts, and from Moſcow 1800 verſts.

Kiſlar,

Kiflar, ... on the Terek, 240 verfts from the above.
Mofdok, ... Ditto 34 verfts from ditto
Georghiyevfk, on the Malaya Kuma, 60 verfts from ditto.
Stauropole, ... on the Afhla, 224 verfts from ditto.

The foil in the fouthern part of the province of Caucafus is very rich and fruitful, but particularly fo on the banks of the Terek, and in the vicinity of Mofdok. There are feveral fine forefts about the Terek, Malka and Kuma, and the Malaya. *Plescheef.*

Page 38. In regard to foffil bones, Pallas fays: "Throughout Afiatic Ruffia there is no river or ftream, particularly thofe which flow through the plains, in the banks and beds of which, bones of elephants and other animals, foreign to the climate, have not been found. The more elevated regions, the primitive and fchiftofe chains, do not furnifh marine petrifactions; whilft the lower declivities, and the large muddy and fandy plains, abound with them, wherever the foil has been wafhed away." In general, it is fingular that foffil bones are not difcovered in climates now inhabited by elephants, whilft they are fo common in latitudes where thofe animals cannot live at prefent.

Page 52. The famous caverns near Inkerman in the Taurid, appear to be of a fimilar nature with thofe of Dufched. "The rock is fo completely excavated, that its open cells, which, like ftories of a houfe, are divided into rows, refemble at a diftance, in fome directions, the form of bee-hives. The church has been formed out of a fingle piece of this rock. Sarcophagi, of coarfe workmanfhip, are frequently found in the cells that communicate with each other. There is a ftair-cafe, leading from the church, upwards to the level ground occupied by the fort."—*Pallas.* Our town of Nottingham, likewife, furnifhes caves anciently dug in the rocks, where the inhabitants keep the vaft quantities of malt liquors made there.

Page 64. The Kumuk Tatars are divided into fix principal tribes. *Plescheef.*

Page

NOTES.

PAGE 65. The Nogay or Mankat Tatars are confidered as the fecond main branch of the nation. Of this powerful race we have hitherto no accurate and connected accounts. According to the Arabian and Grecian hiftorians, they defcended from a Mongul commander of the name of Nogay; who, at the clofe of the thirteenth century, was difpatched by a Khan of Kaptfchak, with a ftrong force, to take poffeffion of the countries bordering on the Black Sea. He, in fact, fubjugated the countries from the Don to the Donau, feized afterwards the Khanfhip of Kaptfchak, and founded an independent ftate; which, however, was divided amongft his fucceffors. Notwithftanding the fall of the ftate, the name of the founder lives in the nation. It confifts of numerous large and fmall tribes, who frequently change their refidence and their names. *Storch.*

IBID. The Daghefttanic Tatars have a cuftom which they carefully obferve; namely, that no perfon can marry, before he has planted, in a particular fpot, one hundred fruit trees; fo that we find, all over the mountains, large forefts of them. *Hist. of the Tatars.*

PAGE 72. The trade of Derbend is very trifling, there being very little demand for goods amongft the neighbouring Caucafian nations, who are moftly poor. And, except a fmall quantity of faffron, wine, filk, cotton, and rice, Derbend has no produce to export. In Gmelin's time, two or three veffels only went thither from Aftrachan, with iron, fteel, lead, and fome rye and oats. The whole trade did not amount to much more than 150,000 roubles, and has decreafed fince that period. Derbend was taken by the Ruffians on the 10th of May, 1796. The news of the capture caufed very great fenfations at Peterfburg; it was announced by firing of guns and ringing of bells. All the perfons attached to the petty court of Count Zonbow, the commander in chief, were remunerated. *Storch. Masson.*

PAGE 85. Marfhal Bieberftein mentioning that he remained ſome time at Kurt-Bulak, our readers may not know, perhaps, that he was attached to the unfortunate expedition againſt Perſia. We ſhall introduce a view of it. After the occupation of Derbend, Baku, and Schamachi, the Ruſſian army was obliged to halt. The rainy ſeaſon, accompanied with exceſſive heat, which the Ruſſian ſoldier can ſcarcely ever bear; the immoderate uſe of fruit and melons; peſtitial winds; and the natural inſalubrity of the country; ſuſpended their operations. Contagion began to ſpread through the troops, and the reinforcements which continually arrived, ſerved only to increaſe the ſick liſt, and add to the general diſtreſs. Being obliged to retire to the mountains, where the air is leſs injurious, the army was conſtantly haraſſed by the mountaineers. This petty warfare coſt the Ruſſians a great many men. In ſhort, in October the mountaineers, ſupported by a body of the Perſian cavalry, made a grand attack; they ſurpriſed ſome battalions of chaſſeurs, and the battle became general. The enemy loſt above one thouſand men; the Ruſſians ſuffered very conſiderably, and had to regret many brave officers. Thus, one of the fineſt armies that ambition ever ſacrificed, was nearly annihilated. This campaign may be ranked, one day, in hiſtory, as a companion to the French expedition to Egypt, of which it appears to be a parody. The motives were nearly ſimilar, and it may have the ſame influence on the reſt of the world.

Masson.

Page 88. Baku imports from Aſtrachan, cloths, drugs for dyeing, iron, linens, and mercery; and exports ſaffron, cotton, rice, ſilk, wine, petroleum, opium, and fruits. Six or eight veſſels arrive annually from Aſtrachan, the cargoes of which are eſtimated at between 300,000 and 400,000 roubles. *Storch.*

Page 89. We may calculate at 8000 *pouds* annually the quantity of ſilk, moſtly from Gilan, imported for the Ruſſian manufactories. The beſt ſort coſts 200 roubles, that of inferior quality 150 roubles, per *poud.* That of Schamachi varies

varies in price from 135 to 170 roubles. Though this article is duty free, yet, in the entries, a falſe ſtatement of the quantity and value is generally made. As the exportation from Aſtrachan never equals the importation, and the ſilk is bought rather cheaper for ready money, moſt of the ſpecie goes out of Ruſſia for that article alone. Certainly we do not reckon the ſum too high by ſtating it at one million of ſpecie, in ducats and roubles, ſent annually to Perſia for ſilks. *Storch.*

PAGE 91. Nadir Shah fell, at the age of 60, by the hands of four of his generals. From an humble ſtation he had raiſed himſelf to a degree of power, at which few monarchs by birth have ever arrived. He ſeems to have united the talents of a complete general and an-an able politician; and though he had not the advantages of learning, yet appears to have had a taſte for true magnificence; and would, probably, had he lived in happier times, have encouraged the arts of peace, and been no ſtranger to the charms of ſociety. But the darling object of his life, to which he ſacrificed every other purſuit, and devoted all the powers of his mind and body, was the art of war; in which he became equal to the greateſt commanders of Aſia, and may juſtly ſtand upon a level with Cyrus or Tamerlane. He had neither the raſhneſs of Alexander, the inſatiable ambition of Cæſar, the inflexible obſtinacy of Charles XII. nor the vices of his illuſtrious rival Pompey the Great. He reſembled, rather, that real hero, Guſtavus Vaſa, who, to uſe the words of Voltaire, "left the foreſt where he lay concealed, and came to deliver his country." Like Vaſa, he was raiſed to the throne of the empire which he had freed from oppreſſion; like Vaſa, he changed the religion of his ſubjects; but he did not, like Vaſa, reign happy and beloved to an advanced old age.

Jones.

PAGE 94. Scythia ſeems to be the general name, which the ancient Europeans gave to as much as they knew of the country. But, whether that word be derived, as Pliny ſeems to intimate, from *Sacai*, a people known by a ſimilar name to the Greeks and Perſians; or, as Bryant imagines, from *Cuthia;* or, as Colonel Vallancey believes, from words denoting *navigation;* or, as it might have been ſuppoſed. from a Greek root, implying *wrath* and ferocity: this, at leaſt, is certain: that neither Scythia nor Tartary are names by which the inhabitants of the country have ever diſtinguiſhed themſelves. *Jones.*

Page 109. Pliny thus deſcribes the *Portæ Caspiæ:*— "Cauſa portarum nominis eadem qua ſupra, interruptis anguſto tranſitu jugis, ita ut vix ſingula meent plauſtra, longitudine viii. m. paſs. toto opere manufacto. Dexterâ lævâque ambuſtis ſimiles impendent ſcopuli. Portæ Caucaſiæ, magno errore à multis Caſpiæ dictæ, ingens naturæ opus montibus interruptis repentè, ubi fores obditæ ferratis trabibus, ſubter medias amne diri odoris fluente, citraque in rupe caſtello (quod vocatur Cumania) communito ad arcendas tranſitu gentes innumeras."—With regard to the firſt and ſecond paſs, in the northern chain of Caucaſus and near Derbend, there appears to be no doubt; we can only aſk then, whether the Marſhal has really found a third paſs in the mountains, behind Derbend; and what grounds he has for believing, that it is one of thoſe which we meet with among the ancients? Otherwiſe, it is more probable that the third paſs, commonly called *Via Caspia*, lay much more to the South than Derbend; becauſe it muſt be the one which Alexander paſſed through, who never was in the neighbourhood of Derbend. *Zach.*

Ibid. The poſition of Moſdok on the Terek, and as the central point of the Ruſſian frontiers, is certainly very advantageous for commerce. But the wild neighbouring nations are great impediments to its extenſion. Moſdok is moſt important as a frontier garriſon; as a place of trade it is inſignificant. From 1793 to 1797 the imports amounted to only 2,085 roubles, an the exports to 1,847 roubles. Moſdok is 130 verſts from the firſt Georgian village of Stepan

Zminda,

Zminda, and 252 from Teflis. The determination of Catherine II. to make a high road from Mofdok to Teflis, had it been carried into execution, would not only have protected the trade, but been a check to the inroads of the plundering tribes of Caucafus. They are too poor in general, and too little acquainted with artificial wants, to promote trade to any great extent. Tin, lead, fteel, iron articles, Ruffia leather, and gunpowder, are in the greateft requeft among them; for thefe they barter wax, honey, fkins, horfes, and female flaves. Total of the Ruffian trade on the Caucafian frontiers, from 1793 to 1797: imports 93,704 roubles; exports 54,753 roubles. *Storch.*

PAGE 112. Until the reign of Peter I. the Ruffians began their year in September, and dated their æra from the creation of the world. In 1700, Peter inftituted a grand jubilee at Mofcow, and ordained that from that period the year fhould begin in January, and be computed from the Chriftian æra, according to the old ftyle then in ufe in England. Out of veneration to his memory, no alteration has fince been made in the Ruffian calendar. *Coxe.*

PAGE 115. The Caucafian mountains in India, according to Captain Wilford, derived their name from the Khafa, a tribe of people whofe defcendants ftill inhabit that region of the earth. Their original country was Cafhgar, to the N E. of Cabul, fituated in a beautiful valley, and watered by a large river, which runs through Chatraul, its capital city. But the name of Caucafus extended from India to the Mediterranean, from the circumftances of the Khafa having once inhabited the vaft range of mountains which lies between thofe diftant points. A principal part of the Caucafus, called *Paramisus* by the Grecian geographers, appears to have been derived from the Sanfcrit words *para vami*, or the pure city of Vami; the fame place which, fince the Mahommedan conqueft, has been commonly called Bamiyan. This city is fituated on the road between Balkh (the ancient Bactria) and Cabul.

PAGE

PAGE 117. Profeſſor Pallas had ſome ſpecimens of granite and granitell, ſent to him by Count Muſſin Puſhkin, from the *apex* of Mount Beſhtau, which leaves no doubt, ſays he, that this inſulated part of Caucaſus contains a rocky grain of the granite kind, on the ſurface of which is depoſited the calcareous matter uſually found in mountains. The Profeſſor adds: "From a laudable zeal of exploring mineralogical objects, the noble Count has, in the preceding year, (1800,) at his own expence, undertaken a journey into the lofty regions of Caucaſus; and we may juſtly expect many important elucidations of this remarkable chain of mountains."

Hunter's Translation.

PAGE 120. The Tchouvaches and Tatars gargle their throats with aſphaltum when they have ulcers. They alſo apply it to freſh wounds; and mix it up with butter, as an ointment for all kinds of ulcers. They alſo boil a ſpoonful of it in milk, and take it as a remedy for the cholic, and pains in the bowels. *Pallas.*

Aſphaltum, mixed with other matters, makes an excellent cement, incorruptible by air, and impenetrable by water. This was ſuppoſed to be the mortar ſo much celebrated among the ancients, wherewith the walls of Babylon, and the temple of Jeruſalem, were cemented. It yields an oil which defends ſhips from water, worms, &c. much better than the ordinary compoſition. *Encyclopædia.*

PAGE 122. Earth impregnated with fire is not peculiar to the neighbourhood of the Caſpian, for we meet with it in the Eaſt Indies. Sir W. Jones's noble biographer thus deſcribes the burning well near Chatigan in the province Chittigong. The ſhape of the well is oblong, about 6 feet by 4, and the depth does not exceed 12 feet. The water, which is always cold, is ſupplied by a ſpring, and there is a conduit for carrying off the ſuperfluity. A part of the ſurface of the well, about a fourth, is covered with brick work, which is nearly ignited by the flames, which flaſh without intermiſſion from the ſurface of the water. It would appear that an inflammable vapour eſcapes through the water, which takes fire

on

on contact with the external air. The perpetuity of the flame is occasioned by the ignited brick-work, as, without this, much of the vapour would escape without conflagration. This was proved by taking away the covering of brick-work, after the extinction of the heat, by throwing upon it the water of the well. The flames still continued to burst forth from the surface, but with momentary intermissions; and the vapour was always immediately kindled by holding a candle at a small distance from the surface of the water. A piece of silver, placed in the conduit, was discoloured in a few minutes; and an infusion of tea gave a dark tinge to the water. On the side of a hill, distant about three miles from the well, there is a spot of ground, of a few feet in dimensions, from which flashes of fire burst on stamping strongly with the foot. The appearance of this spot resembled that of earth on which a fire had been kindled."

Page 123. There is a kind of conjectural geography, which endeavours to ascertain the ancient state of places from modern observations. The French traveller, Olivier, has examined the country, with a view of investigating the truth of the communication, which was said to have existed formerly, between the Euxine and Caspian seas. He is of opinion, that it took place to the northward of Caucasus; and that it has been interrupted by the alluvions of the Kuban, the Volga, and the Don. Since that period, the Caspian not receiving a sufficient volume of water, from the rivers that flow into it, to supply the evaporation, has gradually lost, and at present is sixty feet lower than the Euxine. By the same means it became detached from the lake of Aral, and has receded from those immense sandy and saline plains which surround it to the north and east.

Page 158. Amongst the various improvements lately made in regard to food for animals, the following method of feeding sheep may be new to most of our readers. In Saxony they carefully gather the fruit of the Indian chesnut, which is considered as a very nutritious food, and as a specific for the rot. It is given to sheep in autumn, when green fodder can

no

no longer be procured. The chesnuts are cut into pieces, and given in the proportion of five quarters of a pound to each sheep. The proportion for lambs is not so great. The sheep and cows at first refuse the chesnuts, but, when once habituated, they become very fond of them, and will even eat them with the skins on. However, it is dangerous to give them whole, as the animals might be choked. *Lasteyrie.*

PAGE 159. The most powerful nations, as well as the largest rivers, have been insignificant in their origin. Nature, which made vultures and doves, creates also that ferocious band, that is one day to rush upon the peaceful society which has been formed in its neighbourhood, or which it may meet with in its wandering incursions. The purity of blood among nations, if we may be allowed the expression, as well as the purity of blood among families, cannot be more than temporary, unless kept up by whimsical or religious institutions. A mixture is the necessary result of an infinite number of causes; and from this mixture a race universally springs up, which is either improved or degenerated, according as the character and manners of the conqueror, or as the character and manners of the conquered have given way to those of the conqueror. Among the various causes which soonest bring about this intermixture, that which presents itself as the primary and principal one, is emigration; more or less excited by the barrenness of the soil, and the disagreeableness of the residence. *Raynal.*

Understanding that Gueldenstædt's Travels are not easily procured in this country, we shall close these notes with his catalogue of plants indigenous to the Caucasian isthmus. "On the banks of the Terek," says he, "and in the adjacent countries, I have observed the following plants."

Blitum virgatum.
Hippuris vulgaris.
Callitriche verna.

Veronica

NOTES.

Veronica agreſtis.
——— triphyllos.
——— ſpicata.
——— officinalis.
——— multifida.
——— longifolia.
Veronica campeſtris.
——— nutans.
——— Æthiops.
——— Nemorenſis.
Liguſtrum vulgare.
Circec lutetiana.

Crocus vernus.
Valeriana phu.
——— locuſta olitor.
——— officinalis.
Iris deſertorum.
—— pſeudacorus.
—— pumila.
Gladiolus imbricatus.
Plararis arundinacea.
Hordeum murinum.
Elymus arenaria.
Alopecurus geniculatus.
——— paniculatus.
——— pratenſis.
Cynoſurus durus.
Stipha pennata.
——— capillata.
Bromus mollis.
Alopecurus criſtatus.
Feſtuca vivipara.
Melica nutans.
Lolium perenne.
Avena purpurea.
Holcus odoratus.
Panicum dactylon.
Arundo Calamagroſtis.
Holoſteum umbellatum.

Dipſacus laciniatus.
Valancia cruciata.
Plantago lanceolata.
——— media.
Galium Aparine.
——— rubioides.
——— tinctorum.
Rubia tinctorium.
Potamogeton puſillum.
Potamogeton natans.
——— perfoliatum.
Cornus mas.
——— ſanguinea.
Euonymus Europæus.
Scabioſa Arvenſis.
——— ochroleuca.
Elæagrus anguſtifolia.
Cuſcuta Europæa.

Ulmus

Ulmus campeſtris.
——— pumila.
Lithoſpermum arvenſe.
——— purpuraſcens.
Primula uniflora.
——— veris officin.
Myoſotis ſcorpioides.
Cynogloſſum decumbens.
——— album.
——— officinale.
——— Luſitanicum.
Verbaſcum Phœniceum.
——— nigrum.
——— thapſus.
Vinca minor.
Myoſurus minimus.
Symphytum officinale.
Lycopſis pulla.
——— Arvenſis.
Aſperugo procumbens.
Androſace maxima.
Daucus carota.
——— Mauritanicus.
Sium Falcaria.
Tordylium nodoſum.
Cachrys ſicula.
Smyrnium perfoliatum.
Chærophyllum bulboſum.
Anthriſcus odoratiſs.
Bupleurum rotundifol.
Eryngium planum.
——— campeſtre.
Scandix cerefolium.
Chærophyllum temulum.
Pulmonaria officin.
Lonicera caprifolium.
Viburnum opulus.
Staphylea pinnata.
Sambucus nigra.
——— ebulus.
Rhus cotinus.
Heliotropium Europæum.
Echium Italicum.
——— vulgare.
Tamarix Gallica.
Vitis vinifera.
Rhamnus Paliurus.
——— Frangula.
Linum hirſutum.
——— Gallicum.
——— Auſtriacum.
——— catharticum.
Datura ſtramonium.
Samolus valerandi.
Peucedamum officin.
Crithmum campeſtre.
Aſclepias nigra.
Gentiana centaurium.
Lyſimachia ephemerum.
Apocynum Venetum.
Menyanthes nymphoides.
Solanum dulcamara.
Salſola proſtrata.
——— ſylveſtre.
Conium maculatum.
Seſeli patens.
——— elatum.
——— pumilum.
Athamanta cervaria.
Pimpinella ſaxifraga.
Hyoſcyamus niger.
Periploca Græca.
Theſium linophyllum.
Phyſalis alkekengi.
Statice trigona.
Onoſma echioides.

Onaſma

NOTES.

Onoſma ſimpliciſs.
Cerinthe minor.
Cyclamen Europæum.
Campanula glomerata.
———— lilifolia.
———— patula.
———— medium.
———— perſicifolia.
Convolvulus ſepium.
———— repens.
Convolvulus arvenſis.
Salſola villoſa.
——— tragus.
——— nitroſa.
——— fruticoſa.
——— kali.
Anabaſis aphylla.
Chenopodium polyſperm.
———— album.

Convallaria majalis.
———— multiflora.
———— polygonatum.
Scilla amœna.
—— bifolia.
Bulbocodium vernum.
Ornithogalum luteum.
———— Narbonenſe.
———— minimum.
Galanthus nivalis.
Hyacinthus botryoides.
———— amethyſteus.
Hyacinthus comoſus.
Anthericum ramoſum.
Scirpus paluſtris.
Rumex acetoſa.
——— maritimus.
Tulipa ſylveſtris.
——— Geſneriana.
Aſparagus officin.
Allium urſinum.
——— paniculatum.
——— anguloſum.
Frankenia hirſuta.

Œnothera biennis.
Polygonum amphibium.
——— perſicaria (maculata.)
——— avicular.
Epilobium paluſtre.
Butomus umbellatus.
Daphne Cneorum.

Tribulus terreſtris.
Chryſoſplenium alternifol.
Oxalis acetoſella.
Silene conica.
—— Tatarica.
Agroſtema Githago.

Dianthus

Dianthus prolifer.
——— plumarius.
——— Tauaicensis.
Sedum acre.
Zygophyllum fabago.
Dictamnus albus.
Stellaria geniculata.
——— graminea.
——— holostea
Lychnis dioica.
——— viscaria.
Lychnis chalcedonica.
Alsine media.
Cerastium viscosum.
Arenaria serpyllifol.
——— graminifol.
Cucubalus Otites.
——— Behen.
——— Italicus.
——— fœtidus.
Gypsophila altissima.

Asarum Europæum.
Agrimonia Eupatoria.
Sempervivum vermicul.
——— globiferum.
Euphorbia helioscopa.
Euphorbia longifolia.
——— Cyaprissias.
——— Esula.
——— Chamæsyce.
Portulaca dichotoma.

Potentilla leucantha.
——— argentea.
——— supina.
——— Monspeliensis.
——— Verna.
——— Anserina.
Geum urbanum.
Fragaria vesca.
Amygdalus nana.
Prunus spinosa.
Mespilus Germanica.
Pyrus Pyraster.
——— salicifolia.
Pyrus malus sylv.
——— Cydonia.
Prunus domestica
——— cerasus.
——— Armeniaca.
Cratægus oxyacantha.
Spiræa filapendula.
——— crenata.
Rosa canina.
——— pimpinellifolia.
Rubus cæsius.
——— fructicosus.

Chelidonium

Chelidonium corniculatum.
——— ——— magus.
Papaver hybridum.
——— Rhœas.
Ranunculus aquatilis.
——— ficaria.
——— acris.
——— ſceleratus.
——— Aſiaticus.
——— repens.
——— polyanthemus.
——— Illyricus.
——— caſſubicus.
——— lanuginoſus.
——— lingua.
——— falcatus.
Pœonia tennifolia.
Adonis vernalis.
Adonis æſtivalis.
Anemone ranunculoides.
——— Apenina.
——— pulſatilla.
——— patens.
Delphinium elatum.
——— Ajacis.
——— conſolida.
Thalictrum flavum.
Caltha paluſtris.
Clematis recta.
——— integrifolia.
Capparis ſpinoſa.
Spatiotes aloides.
Aliſma plantago.
Nymphæa alba.
Tilia Europæa.

Lamium purpureum.
——— album.
——— maculatum.
——— amplexicaule.
Ajuga reptans.
——— orientalis.
——— pyramidalis.
Glechoma hederacea.
Teucrium chamæpithys.
——— polium album.
——— chamerdrys.
——— ſcordium.
——— Creticum.
Thymus ſerpyllum.
Thymus acinos.
——— pumilus.
Ballote nigra.
Orobanche ramoſa.
——— major.
Phlomis tuberoſa.
——— herba venti.
Prunella laciniata.
Leonurus cardiaca.
——— marrubiaſtrum.
——— Tataricus.
Sentellaria peregrina.
——— galericulata.
——— haſtifolia.

NOTES.

Galeopſis tetrahit.
Melampyrum arvenſe.
Sideritis perfoliata.
Anthirrhinum minus.
———— geniſtifolium.
Pedicularis comoſa.
Stachys paluſtris.
———— Germanica.
Dracocephalum Ruyſch.
Nepeta Ucranica.
Clinopodium vulgare.

Rhinanthus Criſta galli.
Lathræa ſquamaria.
Scrophularia aquatica.
———— bicolor.
Marrubium peregrin.
Betonica officinalis.
Origanum vulgare.
Dodartia orientalis.
Euphraſia odontites.
Verbena officinalis.

Alyſſum campeſtre.
———— incanum.
Cochlearia draba
Lepidium perfoliatum.
———— ruderale.
———— latifolium.
Thlaſpi burſa.
———— arvenſe.
Siſymbrium Sophia.
Hypecoum pendulum.
Myagrum paniculatum.
———— ſativum

Dentaria pentaphyllos.
Eryſimum alliaria.
Crambi Tatarica.
Turritis glabra.
Iberis nudicaulis.
———— amara.
Sinapis nigra.
———— Hiſpanica.
Draba verna.
Iſatis tinctoria.
Cheiranthus eryſimoides.
Heſperis Sibirica.

Geranium cicutarium.
———— ſanguineum.
———— ciconium.
———— puſillum.
———— Romanum.

Alcea ficifolia.
Althea officinalis
———— cannabina.
Malva rotundifolia.
Sida abutilon.

Medicago polymorpha.
———— purpurea

Hedyſarum Onobrychis.
Glicyrrhiza glabra.

Glycyrrhiza hybrida.
——— echinata.
Cytisus hirsutus.
——— supinus.
Orobus vernus.
——— niger.
Vicia lathyroides.
—— lutea.
—— sepium.
—— pedunculata.
—— cracca.
Coronilla varia.
Ononis spinosa mitis.
——— alopecuroides.
Lathyrus aphaca.
——— hirsutus.
Polygala vulgaris.
Fumaria officinalis.
——— intermedia.
——— bulbosa, solida.
Lathyrus pratensis.
Sophora alopecuroides.
Astragalus glicyphyl.

Hypericum perforatum.

Tussilago petasites.
——— alba.
——— farfara.
Leontodon taraxacum.
Tragopodon pratense.
——— orientale.
Scorzonera purpurea.
Achillea millefolium.
Centaurea Cyanus.
——— sicula.
——— humifusa.
Serratula arvensis.
Tanacetum officin.
Lactuca querna.
After acris.
Erigeron Canadense.
Viola adorata.
—— tricolor.
—— lanceolata.
Viola canina.
Prenanthes muralis.
Carduus mollis.
——— lanceolatus.
——— nutans.
——— cyanoides.
Cichorium intybus.
Inula pulicaria.
—— Helenium.
—— salicina.
Onopordum acanthium.
Cnicus spinosissimus.
Chrysanthemum leucanthemum.
——— corymbosum.
Souchus oleraceus.
Xeranthemum aunuum.
Artemisia linearis.
——— abrotanum.

NOTES.

Anthemis cotula.
——— tinctoria.

Echinops ritro.
Chryſocome villoſa.

Ceratocarpus arenarius.
Atriplex pedunculata.
——— patula.
Typha minor.
Carex nigra.
——— nemoroſa.
——— rufa.
——— panicea.
Najas marina.
Mercurialis annua.
Urtica dioica.
Salix triandra.
Humulus lupulus.
Fraxinus excelſior.
Salix fragilis.
——— caprea.
——— rubra.
Corylus avellana.

Betula alnus.
Acer campeſtre.
——— platanoides.
——— Tataricum.
Populus nigra.
——— alba.
Quercus robur.
Morus Tatarica.
Sparganium amoſum.
Ephedra monoſtachya.
Carpinus betulus.
Fagus ſylvatica.
——— caſtanea.
Xanthium ſtrumarium.
Taxus baccata.
Buxus ſempervirens.
Hydrocharis morus rhanæ.
Viſcum album.

Ophrys bifolia.
——— nidus avis.
Orchis coryophora.
——— conopſea.

Serapias helleborine.
Ariſtolochia clematit.
Arum maculatum.

Equiſetum arvenſe.
——— fluviatile.
Ophiogloſſum vulgare.
Adianthum capill. ven.
Polypod. Filix mas.

Polypod. Filix femina.
——— vulgare.
Aſplenium rutamur.
——— ceterach.
——— adianthum nig.

Oſmunda

Ofmunda ftruthiopteris.
——— trichomanes.
Marfilea natans.
Polytrichum commune.
Conferva reticulata.
Lichen caninus.
——— candelarius.
——— ciliaris.
——— plicatus.
Jungermænia epiphylla.
Boletus igniarius.
Boletus verficolor.
Peziza fcarlatina.
——— punctata.
——— fcuteliata.
Elvela purpurea.
——— pineti.
Phallus impudicus.
——— efculentus.
Tremella noftoc.
Agaricus pectunculus.

Plants growing on the anterior Mountains between Teflis and the River Alasan; in Bloom in February.

Veronica agreftis.
Bulbocodium vernum.
Crocus vernus.
Lamium amplexicaule.
Viola odorata.
Fumaria bulbofa.
Galanthus nivalis.
Scilla amœna.
Ornithogalum luteum.
Hypecoum pendulum.
Sifymbrium Sophia.
Hyacinthus racemofus.
Primula acaulis.
Tuffilago farfara.
Corylus avellana.
Betula alnus.
Ulmus campeftris.
Populus alba.
Vifcum album.
Buxus arborefcens.
Iberis nudicaulis.
Iris pumila.
Vinca minor.
Potentilla verna.
Leontodon taraxacum.
Cornus mas.
Amygdalus communis.
Prunus domeftica.
——— Armeniaca.

At the beginning of March.

Pyrus communis.
——— Cydonia.
Granatus.
Ficus carica.
Fagus caftanea.
Vitis vinifera.

Ranunculus Ficaria.
Lamium album.
Glechoma hederacea.
Scilla bifolia.
Ornithogalum uniflorum.
Aſplenium ceterach.
Aidanthum capil. ven.
Euphorbia helioſcopa.
Meſpilus communis.
Juglans regia.
Thlaſpi burſa.
Geranium Romanum.
Scrophularia canina.
Ornithogalum comoſum.
Ulmus pumila.
Betula alnus.
Populus tremula.
Aſplenium adianthum nig.
Viola tricolor.
Lithoſpermum arvenſe.
Jungermænia epiphylla.
Iris Caucaſia et Georgia.

To the Middle of March.

Rubus fruticoſus.
Roſa canina.
Sambucus ebulus.
Rhamnus paliurus
Prunus ſpinoſa.
Aſtragalus tragacantha.
Quercus robur.
Carpinus betulus.
Fagus ſylvatica.
Cratægus oxyacantha.
Pyrus malus.
Amygdalus Perſica.
Dentaria pentaphyllos.
Eryſimum alliaria.
Simphytum conſolida.
Ajuga pyramidalis.
Hedera on the Carp. Bet.
Anemone cærulea.
Vicia lutea.
Dioſpyros lotus.
Valantia cruciata.

At the latter End of March.

Lathræa ſquamaria.
Cynogloſſum vulgare.
Borago officinalis.
Eryſimum alliaria.
Anemone alba.

In Bloom are included:

Sambucus nigra et ebulus.
Evonymus quadrigonus.
Oſmunda ſtrutiopteris.
Lupuli humuli turiones.

On the eastern anterior Mountains, in Woods on the Alasan.

Cornus mas.
Rhamnus paliurus et ſaxatilis.
Jaſminum humile.
Ficus carica.
Rhus cotinus.
Vinca minor.
Ajuga repens.
Aſplenium ruta muraria.
Hyacinthus botryoides major.
Scrophularia canina.
Tragopogon orientale.
Thlaſpi arvenſe.
Dictamnus fraxinella.
Sedum majus.
Polium album.
Helleborus viridis.
Dentaria enneaphyllos.
Cardamine pratenſis.
Ornithogalum comoſum.

Not in Bloom.

Inula Helenium.
Dipſacus laciniatus.
Hyoſciamus niger.
Primula veris.
Petaſites alba.
Cyclamen Europæum.
Polypodium officinale.
Viola tricolor.
Fragaria veſca.
Orchis Hall. 1279.
Pteris nudicaulis.

In April, near the River Ksani.

Ranunculus polyanth.
Lycopſis arvenſis.
Staphylea pinnatifida.
Cytiſus hirſutus.
Liguſtrum vulgare.
Hippophaë rhamnoides.
Potentilla verna.
Viola elata.
Lithoſpermum purpur.
Ariſtolochia quadrifl.
Pœonia officinalis.
Ajuga reptans.
Cornus ſanguinea.
Lonicera Caprifolium.
Allium urſinum.
Polygonum multiflor.
Euphorbia helioſcopa.
Teucrium chamæpithys.
Berberis vulgaris.
Alſine media.
Cochlearia draba.
Convallaria majalis.
——— multiflora.

Achillea millefolium.
Cichorium intibus.
Mentha Arvenſis.
Salvia verticillata.
Echium vulgare.
Urtica urens et dioica.
Plantago media.
Artemiſia abſinthium.
Rumex aquatica.
Acer campeſtre.
Daphne mezereum.
Cneorum tricoccon.
Thlapſi Arvenſe.
Rhododendron Caucaſ.
Anemone pulſatilla.
Alchemilla vulgaris.
Orchis militans.
Ophrys inſectifera.
Adonis æſtivalis.
Fraxinus excelſior.
Ulmus campeſtris et pumila.
Cratæg. torminalis.
Pyrus pyraſter.
Rhus cotinus.

On the Banks of the River.

Salix alba.
Betula alnus.
Populus tremula et nigra.
Hippophaë rhamnoides.
Elægrus anguſtifolia.
Ophrys nidus avis.
Rumex acetoſa.
Aſparagus officinalis.
Tragopogon Pratenſe.
Vicia ſepium.
Anchuſa officinalis.
Papaver rhœas et argemone.
Simphytum officin.
Myoſotis ſcorpioides.
Lycopſis Arvenſis.
Bromus tectorum.
Hordeum murinum.
Senecio Jacobæus.
Hieracium umbellatum.
Viola tricolor.
Veronica multifida.
Fumaria officinalis.
Salſola vermiculata.
Abſinthium Ponticum.
Chryſocome villoſa.
Polium album.
Thymus ſerpyllum.
Sium falcaria.
Ceraſtium viſcoſum.
Lepidium perfoliatum.
Dictamnus albus.
Jaſminum luteum.
Linum catharticum.
Hedyſarum onobrychis.
Rhamnus ſexatilis.
——— paliuris.
Juniperus ſabina.
Ephedra monoſtachya.
Spiræa crenata.
Verbaſcum Phœniceum.
Vinca minor.
Echium Italicum.
Reſeda odorata.
Scutellaria orientalis.
Orobanche major.
Lathræa phelipea.

Celſia

Celsia orientalis.
Polygala vulgaris.
Seseli pumilum.
Cynoglossum decumbens.

In May, on the Banks of the Kurr, near Teflis.

Jasminum fruticans.
Pyrus Caucasi.
Lathyrus aphaca.
Colutea arborescens.
Achillea tomentosa.
Phlomis herbaventi.
——— laciniata.
——— tuberosa.
Vicia bythynica.
Lychnis dioica.
Juniperus communis.
Phleum Pratense.
Dactylis glomerata.
Valeriana officinalis.
Spiræa filipendula.
Sedum telephium max.
Peganum hermala.
Polygonum aviculare.
Malva rotundifol.

On the Banks of the Ksia, and at the Foot of the Araratian Mountains.

Morus alba et nigra.
Tamarix Gallica.
Salsolæ variæ.
Atriplices variæ.
Atripl. pedunculata.
Chenopod. maritimum.
Sophora alopecuroides.
Capparis spinosa.
Zygophyllum fabago.
Viburnum lantana.
Acer platanoides.
Tilia Europæa.

May Plants on the Ksia.

Lonicera cærul.
Jasminum frutescens.
Rosa eglanteria.
Agrimonia eupatorium.
Aristolochia hirta.
Rhamnus cathartiens.
Convolvulus terrestris.
Orchis pyramidalis.
Ægilops cylindrica.
Scutellaria orientalis.
Teucrum orientale.
——— chamædrys.
——— chemæpithys.
Valeriana locusta.
Polygala vulgaris.
Sylvia nemerosa.
——— verticillata.
——— glutinosa.
Gladiolus imbricatus.
Phleum pratense.

Ornithogalum

Ornithogalum Narbonen.
Origanum vulgare.
Alcea ficifolia.
Malva erecta.
Hedyſarum onobrychis.
Rhynanthus orientalis.
Anthirrhinum cymbalar.
Turritis glabra.
Phyſalis alkekengi.
Bryonia vulgaris.
Bupleurum rotundifol.
Impatiens nolitangere.
Smyrnium perfoliatum.
Aſclepias nigra.
Delphinium Ajacis.
Stachys ſylvatica.
Melampyrum cærul.
Orchis coriophora.
——— bifolia.
Melica nutans.
Hieracium umbellat.
Pedicularis comoſa.
Trifolium purpureum.
——— pratenſe.

On the Ksia and Liachvi, in Jugo subalpino.

Azalea Pontica.
Rhododendron Pontic.
Sorbus aucuparia.
Betula alba.
Ribes rubra.
Polygonum biſtorta.
Veratrum nigrum.
Aſtranthiæ.
Rhinantus vulgaris.
Alchemilla vulgaris.
Clinopodium grandiflor.
Prunella alba et cærul.
Lilium martagon.
Comarum paluſtre.
Tormentilla erect.
Geum rivale.
Carex Cyperoides.
Eriophorum polyſtach.
Lolium temulentum.
Secale Creticum.

On the River Rion and in Imeritia. July.

Rhus coriaria.
——— cotinus.
Quercus robur.
Carpinus betulus.
Betula alnus.
Cratægus oxyacantha.
Juniperus communis.
Cytiſus hirſutus.
Fraxinus excelſior.
Populus nigra.
——— tremula.
Roſa canina.
Hyſſopus officin.
Cornus mas.
Salix alba.
Hippophaë rham.

NOTES.

In Bloom.

Anagallis Arvenſis.
Lyſimachia ephemerum.
Gentiana centaureum.
Cicorium intybus.
Datura ſtramonium.
Onopordon acanthium.
Eringium amethyſteum.
Echinops ritro.
Prenanthes muralis.
Digitalis lutea.
Lolium perenne.
Viburnum opuli var. baccis compreſſis.
Portulaca oleracea.
Philadelphus coronar.
Cratægus torminalis.
Circinalis Gled.
Tilia Europ.
Cornus ſanguinea.
Acer campeſtre et platanoides.
Nymphæa lutea.
Galium luteum.
Prunella officinalis.
Alchemilla vulg.
Verbena officin.
Lythrum virgatum.
Gentiana cruciata.
Lyſimachia vulgaris.
Xeranthemum annuum.
Melica nutans.
Cynoſurus criſtatus.
——— geniculatus.
Mentha ſylveſtris.
Lycopus vulgaris.
Primula veris.
Pinus ſylveſtris.
——— abies.
Taxus baccata.
Ilex aquifolium.
Pyrus pyraſter.
Lonicera caprifolium.
Ciſtus helianthem.
Polygala vulgaris.
Melampyrum cæruleum.
Clinopodium vulgare.
Daucus Mauritanicus.
Symphytum officin.
Chærophyllum ſylveſt.
Caleirana officin.
Clinpodium grandiflor. vulgare
Polygonum biſtorta.
Myoſotis ſcorpioides.
Teucrium chamædrys.
Salvia verticillata.
Pyrola ſecunda.
Gentiana aſclepiades.
Fagus ſylvatica.
Cratægus aria.
Prunus Luſitanica.
Viburnum acerifol.
Acer pſeudoplatan.
Azalea Pontica.
Vaccinium areto-ſtaphyl.
Buxus ſempervirens.
Rhamnus alpinus.
Lilium martagon.
Convallaria verticillat.
Euphraſia officinal.
Epilobium hirſutum.
Grandiflorum.

Heracleum

Heracleum panaces.
Aſtrantia major.
Ranunculus acris.
Centaurea ochroleuca alp.
Trifolium ochroleucum, magnum.
Veronica Gadenſis.
Geranium grandiflor.

On the 21st of July.

Cratægus aria.
Viburnum lantana.
Salix caprea.
Juniperus communis.
Leonurus cardiaca.
Urtica dioica.
Thlaſpi Burſa.
Galeopſis tetrahit.
Veronica Beccabunga.
Prunus Luſitanica.
Rhododendron Pontic.
Vaccinium vitis idea.
——— myrtillus.
Ophrys cordata.
Oxalis acetoſella.
Spiræa aruncus.
Saxifraga rotundifol.
Cerinthe minor.
Rhinanthus elephas var B.
Orchis pyramidata alba.
Orchis coryophora.
Stachys ſylvatica.
Mentha ſylveſtris.
Convolvulus ſepium.
Ononis ſpinoſa.
Althæa officin.
Trifolium melilothus.
Arctium lappa.
Xanthium ſpinoſum.
Hyoſcyamus niger.
Hedera helix.
Rhamnus frangula.
Daucus carota.
Eupatorium cannab.
Clemmatis flammula.
Solanum dulcamara.
Hieracium umbellat.
Pœonia officin.
Pari quadrifolia.

On the Snow-Mountain.

Sagina.
Cymbalaria.
Hypericum orient.
Azalea Pontica.
Ruſcus paraſit.
Rhododendron Pont.
Tormentilla erecta.
Anemone alba.
Stellaria magna.
Pyrola magna.
Swertia perennis.
Pedicularis rubra.
Caltha paluſtris.
Trollius.
Salix.
Veronica ſpicata.
Galium longiflor.
Alchemilla.

In

NOTES.

In August, near Schartali.

Alifma plantaga.
Ranunculus linguo.
Menyanthes trifol.
Scirpus.
Fagus caftanea.
Rhododendr. Caucaf.
Vaccinium arto-ftaphyl.
Eupatorium cannabium.
Diofpyros lotus.
Polygonum perficaria.
Leonurus cardiaca.
Clathrus cancellatus.
Pteris aquilina.
Panicum viride.
——— filiforma.
——— crus galli.
Juncus vulgaris.

Near Kutais, on the 12th of August.

Rufcus aculeatus.
Datura ftramonium.
Malva erecta fylveft.
Verbena officin.
Polygon. Perfica maculat.
——— aviculare.
Sida abutilon.
Ballote nigra.
Rumex aquaticus.
Sigifbeckia orient.
Mentha julegium.
Eryngium ametyft.
Cyperus fufcus.
Oxalis corniculata.
Cucurbita lagenaria.
Hedera helix.

Growing very common on Walls.

Xanthium fpinofum.
Laurus nobilis.
Panicum viride.
Panicum crus galli.
Azalea Pontica.
Rhododendr. Pont.

In Imeretia, in August.

Daphne.
Ilex aquifolium.
Prunus Lufitan.
Rufeus hypophyl.
Smilax excelfus.
Vaccinium arctoftaphylus.
Philadelphus coronarius,
Colchicum autumnale.
Buxus fempervirens.
Sorbus torminalis.
Mefpilus communis.
Datifca cannabina.

Sigefbeckia

Sigefbeckia oriental.
Eryngium amethyft.
Teucrium Hircanic.
Thalictrum flavum.
Agrimonia eupator.
Ophrys fpiralis.
Ilex aculeatus.
Rhamnus ulmoides.
Fragaria vefca.
Origanum heracleot.
——— vulgare.
Verbena officinal.
Urtica dioica.
Panicum viride.
Prenanthes muralis.
Sida abutilon.
Solanum nigrum.
Verbafcum Thapfus.
Alcea ficifolia.
Hibifcus.
Erigeron acre.
Artemifia vulg.
Anthirrhinum linar.
Plantago media.
Hedera helix.

On the Tscherimela or Cross-Mountains. In September.

Carpinus Betulus.
Pinus fylveftris.
——— picea.
Betula alnus.
Corylus avelana.
Populus tremula.
Azalea Pontica.
Rhododendr. Pont.
Ilex aquifolium.
Cornus fanguinea.
Vifcum album.
Dictamnus fraxin.
Marrubium album.
Tuffilago farfara.
——— petafites.
Typha anguftfol.
Lithofperm. offic.
Rubia tinctorum.
Convallaria polygon.
Echium vulgare.
Equifetum fylvat.
Pulmonaria offic.
Lemna minor.
Colchicum autumn.
Verbafcum thapfus.
Stellana nemorum.
Cynogloffum offic.
Smilax afpera.
——— lævis.
Humulus lupulus.
Sambucus niger.
——— ebulus
Vitis vinifera.
Polypodium vulg.
Tragopogon orient.
Ornithogolum luteum.
Vicia fepium.
Arabis Thaliana.
Anemone pulfatilla.
Cheldonium majus.
Dipfacus laciniat.
Artemifia abfinth.
Cyperus fufcus.
Afarum Europ.

Crocus

NOTES.

Crocus fativus.
Inula Helenium.
Juniperus fabina.
Platanus orient.
Populus alba.
Ulmus campeft.
Arundo phragmit.
Liguftrum vulg.
Evonymus Europ.
Morus alba.
Afparagus officin.
Plantago media.
Sonchus afper.
Leontodon tarax.
Orobanche major.
Staphylea pinnat.
Nymphæa alba ad Dufchet.
Clematis vitalba.
Daphne Pontica.
Anthemis.
Anchufa.
Lonicera alpina.
Pyrus pyrafter.
Malus fylvatica.
Rinanthus elephas.
Geum rivale.
——— urbanum.
Carex pfeudo Cyperus.
Symphitum majus.
Rubus idæus.
Ciftus helianthem.
Clinopodium grandifl.
Stachys alpina.
——— fylvat.
——— Germanica.
Lamium album.
Mentha fylveft.
Spiræa forbifol.
Acer platanoides.
——— pfeudoplat.
——— campeftris.
Achillea ptarmica.
Centaurea ochrol.
Circea major.
Anagallis Arven.
Echinops ritro.
Prenanthes mural.
Lyfimachia ephem.
Angelica fylv.
Borago officin.
Cufcuta Europ.
Polypodium filixmas.
Cnicus oleraceus.
Rhamnus infectorius?
Phyllirea.
Prunus fylvef.
Juncus.

INDEX.

INDEX.

The numeral Letters refer to the Volume, and the Figures to the Page.

E.

F.

G.

Ghyſſr

Keegti,

V. and U.

W.

X.

Y.

Z.

Zeighor,

INDEX.

FINIS.

Printed by W. Darton, and J. & J. Harvey, Gracechurch-street.

Printed in Dunstable, United Kingdom

91220016R00181